Statistical Methods for Communication Studies

A step-by-step guide through various stages of data preparation and analysis, this text demonstrates descriptive and inferential statistics in communication studies.

This book equips communication students with the tools they need to engage in quantitative research, ensuring they understand why the statistical tools they chose are most appropriate for their data. Using examples and data directly relevant to communication researchers, this text provides guidance on how to use and interpret SPSS and write up results, with further guidance via free online material in using Excel and R. Instead of focusing on formulas and in-depth mathematics, this book provides a hands-on approach to understanding the theory, practice, and application of statistical methods in communication. In addition to covering the standard statistical tools used to analyze social science data for hypothesis testing (e.g., correlation, regression, ANOVA), it also covers advanced factor-analytic techniques specific to measurement validation as well as data visualization guidance.

Ideally suited for statistics or quantitative methods courses in communication studies programs at the advanced and graduate level.

Online resources include tutorials on SPSS, Excel, and R for key constructs and practice quizzes for students. Instructors will find an instructor manual and test bank. Please visit www.routledge.com/9781032814872.

Stephen M. Croucher is a professor in the Department of Communication at Clemson University, U.S.A. He studies organizational and intercultural communication. He has authored more than 130 journal articles and authored/co-edited more than 15 books. He has served as the editor of the *Journal of Intercultural Communication Research*, *Speaker & Gavel*, and the *Review of Communication*, and is editor-elect for *Communication Monographs*. He is a fellow of the International Communication Association and president of the World Communication Association.

Stephanie Kelly is a professor of data and communication at North Carolina A&T State University, U.S.A. Her research focuses on presence variables such as immediacy, credibility, and solidarity in workplace, educational, and technology-mediated contexts. She has published over 100 peer-reviewed articles in journals such as *Communication Education*, *Communication Quarterly*, and the *International Journal of Business Communication*. She currently serves as editor-elect of the *Review of Communication*.

Cheng Zeng is an associate professor in the Department of Communication at North Dakota State University, U.S.A. His research focuses on organizational communication phenomena, including organizational dissent, assimilation, and workplace freedom of speech across diverse cultural contexts. His work has been published in well-regarded communication journals such as the *Journal of Communication Management*, *Journal of Strategic Communication*, and *Communication Studies*.

Statistical Methods for Communication Studies

Stephen M. Croucher, Stephanie Kelly, and Cheng Zeng

Routledge
Taylor & Francis Group
NEW YORK AND LONDON

Designed cover image: 3d_kot/Shutterstock

First published 2026
by Routledge
605 Third Avenue, New York, NY 10158

and by Routledge
4 Park Square, Milton Park, Abingdon, Oxon, OX14 4RN

Routledge is an imprint of the Taylor & Francis Group, an informa business

ISBN: 978-1-032-81488-9 (hbk)
ISBN: 978-1-032-81487-2 (pbk)
ISBN: 978-1-003-50011-7 (ebk)

DOI: 10.4324/9781003500117

Typeset in Times New Roman
by Apex CoVantage, LLC

Access the Support Material: www.routledge.com/9781032814872.

Contents

Chapter 1

Introduction to Statistical Methods for Communication

What Will I Learn in This Chapter?

Statistics are like Legos: versatile, well-structured, and each essential for building something meaningful, but only as good as the hands that assemble them. Just as a pile of Legos connected at random creates assemblies of chaos, using statistics without intentional, purposeful care results in misleading conclusions. This chapter explores why statistical tools are useful, and provides an overview of what else you will read in this book. If the Stormtroopers depicted earlier are not warning enough, then let this introduction be a warning that nerdy references will abound.

We're going to be honest with you, not everyone enjoys statistics. One of the authors of this textbook hated statistics as a new doctoral student. In fact, this author's hatred was strong enough to contemplate never doing statistics again. Those thoughts fortunately dissipated when a new

DOI: 10.4324/9781003500117-1

teacher was able to break things down in a better way and explain the building blocks in a more step-by-step approach.

The point of this story is that we all approach statistics in a different way. Some of us are naturally good with numbers, and so statistics come to us more easily. For some, numbers cause anxiety, and so we need a bit more foundation built to prove to ourselves we really do understand. No matter how we approach statistics, we need to realize that statistics are the building blocks of research, particularly in social science. Our primary objective throughout this book is to provide you with the building blocks to excel at statistical methods. Looking to the rest of Chapter 1, this chapter will:

1. Define statistics
2. Discuss the benefits of taking statistics
3. Outline the structure of the book

What Are Statistics?

There are many things that come to mind when people, particularly students, think about the term "statistics." Those thoughts range from "Aaaaaaah!" to recalling past statistical information that has already been of use to you. **Statistics** are tools and techniques for describing, interpreting, and organizing information or data. We encounter statistics everyday: the average test result on a class exam, ERA (earned run average) in baseball, when a weather forecaster says there is a 75% chance of rain today, political polls, standardized tests such as the SAT or ACT, etc. Essentially, statistics help us understand the world around us.

Test Your Understanding 1.1 – What is the definition of statistics?

There are two types of statistics, descriptive and inferential. **Descriptive statistics** describe and organize the characteristics of data. We will talk more about this in Chapter 5 on Describing Data. For example, you might describe the average age of the students in your class (Mean), or the average house price in Houston, Texas (Mean), or the most common hair color in Seattle, Washington (Mode). Each of these describes the data. **Inferential statistics** make inferences (conclusions) from a smaller group of data (a sample) to a larger group (population). We talk more about making inferences in Chapters 2, 5, and 6. For example, you might work for CNN and survey 1000 voters on a proposal to build a new stadium in Town X. Based on these 1000 voters, CNN states 52% of voters in Town X support the building of the stadium. CNN will need to state that their projection has some error in it (which we discuss later). But the key for now is to know that CNN is making an inference from a subset of the larger population.

Test Your Understanding 1.2 – What are the differences between descriptive and inferential statistics?

Benefits of Taking Statistics

So, what are the benefits of taking a statistics class? Why am I here? There are a few benefits to learning statistics. First, and for very practical reasons, it might be required for your degree. Many programs now require students to take a statistics course. Thus, the skills you will learn in this class are those integral to your degree program. Second, it's important to understand how statistics are used around us. Statistical terms and statistics are thrown around all the time. We need to be critical consumers. This course will help you be a more critical consumer of statistics. Once you become comfortable with statistics, that knowledge will become your lie detector for when people talk about data. Third, if you are an undergraduate student planning on going to graduate school, these skills will give you a key foundation you will need to succeed.

Structure of the Book

What follows are 19 chapters that cover varying statistical topics. These chapters are written by three authors with varying backgrounds in statistics and communication. Chapters 2 to 7 introduce you to key concepts in statistical methods. Chapters 8 to 20 cover a wide range of statistical tests. Chapter 2, "Measurement Fundamentals," introduces many of the key statistical terms and constructs that will guide you on your journey through this book. Chapter 3, "SPSS and Excel," describes the SPSS and Excel environments and discusses how to use each in statistical analyses. This chapter serves as a primer for running statistical analyses for later chapters. Chapter 4, "Data Visualization," defines data visualization and describes how to visualize data using SPSS and Excel. Chapter 5, "Describing Data," explains descriptive statistics. The chapter goes into detail on different types of descriptive statistics, and how to use SPSS and Excel to calculate descriptive statistics. Chapter 6, "Normality and Outliers," teaches you all about bias, statistical assumptions, outliers, normality, and how to use SPSS to analyze each. Chapter 7, "The Research Process," covers hypothesis testing, significance, and error.

The second part of the book includes Chapters 8 to 20. Each of these chapters covers a different statistical test. The first part of each chapter describes the statistical test, while the final part of each chapter covers how to run the test. Chapter 8 covers correlations. Chapter 9 provides a description of simple linear regression, while Chapter 10 covers multiple regression. Chapter 11 discusses t-tests (comparing two means). Chapter 12 delves into comparing multiple means, or analysis of variance (ANOVA). Chapter 13 builds off Chapter 12 and covers multivariate analysis of variance (MANOVA). Chapter 14 examines analysis of covariance (ANCOVA) and multivariate analysis of covariance (MANCOVA). Chapter 15 describes repeated measures. Chapter 16 discusses the chi-square (χ^2) test of independence. Chapter 17 defines reliability and relates it to statistical tests and measures. Chapters 18 to 20 discuss validity and measurement building. Chapter 18 describes exploratory factor analysis (EFA). Chapter 19 defines principal components factor analysis (PCA). The final chapter, Chapter 20, examines confirmatory factor analysis (CFA).

As you progress through the book, you will notice the tests will often build off one another. Refer to earlier readings to help you as the statistical tests get more complex. The key to any statistics book, and course, is to take it one step at a time. Each chapter will provide you with the blocks you need to answer any research question or hypothesis you might propose. So, off you go, and let Lego Yoda, Lego Captain America, and their friends take you on a statistical journey.

Chapter 2

Measurement Fundamentals

What Will I Learn in This Chapter?

One of the authors of this book is a huge Star Wars fan. Just like Yoda taught Luke Skywalker the ways of the force, it is our job to teach you the ways of statistics. Before you can truly be a Jedi, you must lift some rocks, do some flips, and face your demons in a cave. Similarly, before you can be a master of statistics for communication sciences, you must understand its key concepts and learn how to talk the talk.

There are countless terms that make up the fundamental vocabulary of statistics. In this chapter, you will be introduced to many of the key constructs that will guide you through your journey. It may appear as though there is a lot of material and "stuff" to learn, and there is. However, in this

DOI: 10.4324/9781003500117-2

chapter, we will set your mind at ease by defining key terms and providing you with a path to success. In this chapter, we will accomplish seven tasks:

1. Discuss the subtle differences between methodology and method
2. Explore the differences between conceptualization and operationalization
3. Delve into the many methods of measurement
4. Describe the different types of variables
5. Define the different levels of measurement
6. Define reliability and validity, and explain the differences between these two statistical terms
7. Define and discuss data sampling

Methodology versus Method

Before this book progresses into key statistical terms, it is important to differentiate between methodology and method. The authors of this book have had many students confuse these terms over the years. A **method** is a specific approach or tool used to collect data. Examples of methods include surveys, focus groups, experiments, case studies, interviews, observations, and many others. **Methodology** is a theoretical framework that explains and supports a method. It is a theoretical perspective that guides your choice of method. Examples of methodological approaches include phenomenology, critical theory, the interpretive approach, positivist social science, or empiricism (which this book emphasizes).

Everyone has a different view of the world, and the question of what makes research, research is something that has been long debated by philosophers such as John Stuart Mill, Karl Marx, Max Weber, David Hume, and many others. What has emerged over the centuries are three primary approaches to research/science: critical, interpretive, and social scientific. Each approach differs on three philosophical assumptions, or positions: axiology, epistemology, and ontology. In this book, we focus on the post-positivist approach to research, which approaches the world from a scientific lens. For discussion on the interpretive or critical approaches, see Croucher and Cronn-Mills (2022) or Neuman (2011).

Social science is one of the three main approaches to methodology. It is an organized form of research that combines inductive and deductive reasoning with empirical testing to test theory and discover causal laws. **Axiology** refers to the extent to which a researcher acknowledges their own values in their writing and research. For a social scientist, like a natural scientist (e.g., chemist), the research process should be as value-less as possible. This means that when conducting research, the presence of the researcher should not significantly impact the results. Interpretive and critical researchers openly acknowledge, to differing extents, their place in the research process. **Ontology** refers to the study of what is real, or the nature of being. For a social scientist, reality can be observed and studied because reality is out there to be identified and measured. For a social scientist, the real world exists independently of people and our interpretation of it; thus, we can study it scientifically. For interpretive and critical researchers, individuals' interpretations and lived experiences influence their interpretations of and contact with the "real world." **Epistemology** is the study of what knowledge is and how it is acquired. This branch of philosophy comes from René Descartes, who called upon philosophers to ask what it is we know about reality. When we acquire knowledge, we can do so using common sense, all the way to the application of scientific methods and controlled tests. For a social scientist, knowledge or the "Truth" is only obtained through scientific methods. For interpretive and critical researchers, commonsense and personal experiences provide various "truths," but for a social scientist, these things do not equate to the "Truth." Throughout this text, you will see how these three philosophical approaches combine to guide the different statistical approaches (methods) you choose to answer your research questions/hypotheses. For a deeper dive into the metatheoretical assumptions of social scientists, see Kelly and Westerman (2024).

Test Your Understanding 2.1 – What is the difference between Method and Methodology?

Conceptualization and Operationalization

As you start on your process of developing a study, you will need to define your key terms and then decide on how you will measure these key terms. As a fledgling social scientist (like Luke was a fledgling Jedi on Dagobah), you need to start at the beginning. You must first define your key concepts. **Conceptualization** is when you take an abstract concept or construct and refine it by giving it a theoretical or conceptual definition. Unless you are creating a new concept, this process has often already been done for us in the literature. However, when conducting a study, it is critical to provide a conceptual definition of all the key terms that you will be using in your study. A **conceptual definition** is a systematic definition of a theoretical construct. This definition is often provided in a review of literature, defining a key construct, and then linking that construct (or theory) to other constructs. For example, in Fall et al. (2013), the authors explored emotional intelligence as a predictor of intercultural communication apprehension among university students in the United States. In their literature review, the authors defined intercultural communication apprehension (ICA) and emotional intelligence (EI). When defining both terms, the authors cited original literature that defined each theoretical construct and then linked each construct to other constructs. The purpose of this literature review was to provide a systematic definition of their key concepts, so that all readers know the concepts the authors are studying. In a second example, Croucher et al. (2009) explored the link between organizational identification and organizational culture within intercollegiate forensics teams. In their literature review (conceptualization), they defined organizational culture and identification, using previously published works, and then linked these two theoretical constructs together. If you look on Google Scholar, or any other search engine for other articles, theses, dissertations, or other documents that examine ICA, EI, organizational culture, and organizational identification, you will find researchers have written reviews of literature that are vastly different. The key is that each document provides enough information on the concept(s) to guide the reader from an abstract understanding to a more systematic understanding of the concept(s).

Once you have a clear understanding of your concepts, theories, constructs, etc., you can then move on to the question of how you will measure them. **Operationalization** is when a conceptual definition is linked with tools of measurement. To facilitate this, you will need to provide an **operational definition** of each of your concepts, or a method by which each conceptual definition will be measured. In Fall et al. (2013), the authors explain in their Method section, under Procedures, that ICA "was measured using Neuliep and McCroskey's (1997) Personal Report of the Intercultural Communication Apprehension measure" (p. 418). They then state EI "was measured using the Trait Emotional Intelligence Questionnaire-Short Form (Petrides & Furnham, 2006)" (p. 418). Both statements directly identify how the authors operationalize their variables. In both cases, the authors are using pre-established measures of each construct. Similarly, in Croucher et al. (2009), the authors stated, "To operationalize organizational identity/identification, we used Cheney's (1983) organizational identity questionnaire (OIQ)" (p. 79). Additionally, to assess culture, the authors stated, "To measure organizational culture, Glaser et al.'s (1987) organizational culture (OC) survey was used" (p. 80). Each article links its concepts to clearly described and justified measures that put those concepts into operational form. The reader is not left asking themselves, "How did they measure X?"

Test Your Understanding 2.2 – What is the difference between a conceptual and operational definition?

Methods of Measurement

So now that you have decided on your concept and know how you will measure it (operationalization), you need to collect data. **Data** is information collected in a systematic manner. Data can be non-numeric or numeric in nature. In communication research, there are typically four types of data: texts, observations, self-reports, and other-reports.

Texts are written, performed, spoken, and/or symbolic messages. Texts can be intentional or unintentional, such as non-verbal messages. If your interest is in written texts, media messages, newspaper articles, and article titles are types of texts. Increasingly, communication researchers are studying the influence social media has on our lives. Sadiq et al. (2023), for example, studied YouTube headlines/titles and comments on YouTube videos to examine causes of COVID-19 vaccine hesitancy in Nigeria. The authors collected YouTube videos uploaded from March 2021 to December 2022 that included terms such as "COVID-19 in Nigeria" and "COVID-19 vaccination in Nigeria." Their textual data were the title/headline and the comments for each video. In total, they collected 319 videos and 985 comments.

A second kind of data is **observations** or watching human behavior in action. In this type of data collection, the researcher observes human behavior, takes notes, and ultimately, those notes and the researcher's observations are turned into data. This type of data is often collected in the natural sciences when scientists observe animals, for example. It is also popular among communication researchers. A seminal work using this kind of data was Philipsen's (1976) analysis of "Teamsterville," a neighborhood in Chicago. Philipsen spent a great deal of time observing people who lived there and how they talked and lived. He learned things such as the importance of the porch as a place to talk and observe people and the ability of residents to "spot" outsiders.

The third kind of data is self-report data. A **self-report** is when a participant is asked to report information about themselves. This information can be about their own behaviors, thoughts, feelings/emotions, etc. This kind of data is typically quantitative in nature (but may be qualitative) and collected using surveys. Quantitative self-reports are normally closed-ended assessments that are either Likert scales or semantic differential scales. A **Likert scale** is used in multi-item measures of attitudes, behaviors, and opinions. A Likert scale has five characteristics that differentiate it from response ranges, which are shown in Figure 2.1.

Many of the measures used in communication research utilize "**Likert-type**" scales, response ranges that do not meet one or more of the Likert scale characteristics. For example, the question might ask, "How often do you watch YouTube?" with a response scale of: Never (1), Sometimes (2), Average (3), Often (4), and Very Often (5). You will notice that the verbal levels are not symmetrical; there is no "exact" opposite to "Never" or "Very Often." While it might seem like we are being pedantic, it is important to know the difference when someone writes or says "Likert" or "Likert-type." There is nothing wrong with a Likert-type scale; it is just different. In the Fall et al. (2013) study, ICA was measured using 14 Likert-type items ranging from 1 to 7, using disagree strongly to strongly agree. The Croucher et al. (2009) study used a Likert scale: 1 (to a very little extent), 2

Characteristic
1. Scale has several response options
2. Response levels arranged horizontally
3. Response levels are anchored with consecutive integers
1, 2, 3, 4, 5, etc.
4. Response levels are anchored with verbal levels that represent evenly spaced steps
Strongly disagree to disagree to neutral to agree to strongly agree
5. Verbal levels are bipolar and symmetrical around a neutral point
Strongly disagree (1), somewhat disagree (2), neutral (3), somewhat disagree (4), and
strongly disagree (5) – are symmetrical and bipolar.

Figure 2.1 Characteristics of a Likert scale.

(to a little extent), 3 to some extent, 4 (to a great extent), and 5 (to a very great extent) to measure organizational culture. The most common change that shifts an item from Likert to Likert-type has been driven by the rise of online questionnaires designed for mobile devices. To fit the screen without requiring scrolling, Likert items are often displayed with response options listed vertically rather than horizontally. This vertical presentation of the scale classifies an item as Likert-type rather than Likert.

A second type of self-report measure often used in communication research is the semantic differential. A **semantic differential scale** uses a response range that asks participants to rate something (concept, idea, brand, person, abstract idea, etc.) on a standard set of bipolar adjectives (opposite meanings). The scale typically is measured on a five or seven-point range. Examples of bipolar adjectives are bad-not bad, unfair-fair, unpleasant-pleasant, dirty-not dirty, unintelligent-intelligent, untrained-trained, stupid-not stupid, dishonest-honest, and unselfish-selfish. Respondents are asked to indicate on a range of blanks between words how they feel when semantic differential scales are used.

Note that in a well-constructed semantic differential scale, the ranges are not always as expected. The opposite of "bad" was not "good," but rather "not bad." In social science measurement, researchers have learned that the opposite of many feelings and values is often not the word commonly thought of as its opposite. Consider your feelings about Brussels sprouts. If someone has never had a Brussels sprout, you might set up a semantic differential item for indicating feelings towards them like this:

Love _ _ _ _ _ _ _ Hate

If these semantics were truly exact opposites, then someone who has never tried Brussels sprouts would have neither love nor hate for them. Yet, this scale forces someone to fall at least midway if they have no experience whatsoever with Brussels sprouts. In our minds, the middle score might seem neutral, but mathematically, it represents 3.5 points towards love and 3.5 points towards hate. This means our measurement has 3.5 points of noise for anyone with no feelings about Brussels sprouts. Instead, the semantic differentials should look like this:

Love _ _ _ _ _ _ _ Not Love
Not Hate _ _ _ _ _ _ _ Hate

Many popular communication measures use semantic differential items. For example, McCroskey and Teven's (1999) Measure of Source Credibility is an 18-item semantic differential scale. The measure assesses the credibility of a person (identified by the researcher) with a 7-point scale; see Figure 2.2 for sample items.

Whether it's Likert, Likert-type, or semantic differential measures, many researchers have asserted that participants over- and underestimate their attitudes, opinions, and behaviors, which is called the social desirability bias (Nicotera, 1996; Podsakoff & Organ, 1986). In fact, it has become common practice for many researchers using self-reports to talk about the potential social desirability bias as a limitation in most published research. However, research from one of the authors of this text (Croucher et al., 2013, 2017) has found that the social desirability bias is often exaggerated and self-reports are valid and reliable. We will talk more about both terms shortly.

The final kind of data is other-report data. **Other-reports** are when you ask individuals to report on the communicative attitudes, opinions, and behaviors of others. For example, let's say you are interested in the extent to which intimate partners express jealousy. You would first write up a conceptual definition of jealousy (and if you were focusing on the expression and experience of jealousy, you might focus on the work of Guerrero and associates). This might lead you to want to use the 70-item Communicative Responses Romantic Jealousy (CRJ) Scale (Guerrero et al., 1995). In doing your research, you are interested in not only self-reported jealousy but how each partner perceives each other. Thus, you have each partner in an intimate relationship complete a self-report

Intelligent	1	2	3	4	5	6	7	Unintelligent
Untrained	1	2	3	4	5	6	7	Trained
Self-centered	1	2	3	4	5	6	7	Not self-centered
Insensitive	1	2	3	4	5	6	7	Sensitive
Honest	1	2	3	4	5	6	7	Dishonest
Informed	1	2	3	4	5	6	7	Uninformed

Figure 2.2 Sample items from the Measure of Source Credibility.

of the CRJ and also an other-report of the CRJ about their partner. A self-report item would say: "When I feel jealous I tend to ignore my partner." While the other report item would say: "When my partner feels jealous he/she tends to ignore me." Both statements would be answered using the Likert-type scale ranging from 1 = strong disagreement to 7 = strong agreement. After you collect your data, you can compare results and see how similar/different the other and self-reports are. Other reports can show you how people perceive themselves versus how others perceive them. Now that we have gone through the different kinds of data, the next section of this chapter outlines the different kinds of variables used in research.

Test Your Understanding 2.3 – What are the four typical kinds of data?

Kinds of Variables

Throughout the chapters in this book, we will be providing you with tools to explore research questions and test hypotheses. A **research question** is a question that explains or explores a claim or assumption. The question is based in literature but is exploratory in nature when compared to a hypothesis. A **hypothesis** is a testable statement, based in previous literature, that expresses how two or more variables are related to one another. In Croucher et al. (2009), they posed two research questions and one hypothesis. "To what extent does competitor's organizational identification differ based on sex?" and "To what extent does competitor's organizational identification differ based on ethnicity?" (Croucher et al., 2009, p. 77). While previous research has demonstrated that identification to differ based on member ethnicity and sex, no research has explored identification among forensics competitors. Thus, the researchers sought to explore these constructs in a new context. The hypothesis asserted: "A competitor's organizational identification with his/her forensics program will be positively correlated with: (a) teamwork, (b) morale, (c) information flow, (d) involvement, (e) supervision, and (f) communication during meetings" (Croucher et al., 2009, p. 79). This was based on years of previous research showing a positive correlation between identification and culture. In exploring research questions and testing hypotheses, variables are measured.

A **variable** is a construct or attribute that can vary or change. For example, if a researcher is studying how much food a cat eats per day, the key variable is the amount of food eaten per day. In any study, researchers will have a dependent and independent variable. The variable we often attribute to causing change in another variable is the **independent variable**, as its value does not depend on the other variables. The variable we are measuring to see if it changes or has an effect due to the influence of the independent variable is the **dependent variable**. For example, Croucher et al. (2022) explored the extent to which patient self-advocacy predicted COVID-19 vaccine non-hesitance. Patient self-advocacy is the independent variable and COVID-19 vaccine non-hesitance level is the dependent variable. In another study, Kim and Oh (2023) predicted immigrants would report higher levels of telephone anxiety. Being an immigrant is the independent variable and telephone anxiety level is the dependent variable.

You may often see "predictor" and "outcome" variable used instead of independent and dependent variable. A **predictor variable** is a variable that predicts an outcome. An **outcome variable** is a variable that changes because of a predictor's influence on it. These two terms (predictor and outcome) will be used more regularly as you advance in this text. Moreover, as you progress throughout this text, you will be introduced to other kinds of variables, particularly when you get to the ANOVA, regression, and factor analysis chapters. For now, focus on understanding predictor/independent and outcome/dependent variables.

Levels of Measurement

In the previous section, we discussed how there are different kinds of variables. In addition to there being different kinds of variables, variables can have different levels of sophistication or complexity. The link between what is being measured and the numbers they represent is the level of measurement. There are two primary levels of measurement, categorical and continuous variables. A **categorical** variable (also sometimes called a "discrete" variable) is composed of categories or distinct things/entities. For example, race/ethnicity is a category. In each nation, the categories differ. For example, according to StatsNZ (2020), individuals in New Zealand can choose one of the following ethnic groups (Category) on the official census: European, Māori, Pacific Peoples, Asian, MELAA (Middle Eastern/Latin American/African), Other Ethnicity, or Not Elsewhere Included. While in the United States, race is divided into the following categories on the Census: White, Hispanic/Latino/Spanish origin, Black/African American, American Indian or Alaskan Native, Asian, Native Hawaiian, other Pacific Islander, or other (Pew Research Center, 2021). There are two kinds of categorical variables, nominal and ordinal variables.

Nominal variables (or data) are categorical variables that have multiple categories but do not have a numeric ordering. Nominal variables also do not overlap. We add data in parentheses because the variables are converted into data when you conduct your analysis. This is discussed in length in the following two chapters. For example, a survey could ask participants to choose their hair color out of a list of six options. None of the colors is ranked higher than the other, no color has a higher value than the other. Each is weighed the same numerically. Participants who choose blonde, red, brown, black, blue, or multi-colored hair are all equal, statistically speaking. The colors are just categories.

The second level of sophistication in measurement is ordinal level data, or variables. An **ordinal variable** is a categorical variable (or data) that is rank ordered. Ordinal variables share the same characteristics as nominal variables, and these variables are ranked in some way: best/worst, highest/lowest, least/most, etc. An example of an ordinal variable is educational level. It is common on surveys to ask participants to identify their highest level of educational achievement: high school – some university – university degree – some post graduate education – post graduate education – doctorate. There is a clear ordering of education with these options. However, the distance between these options is not entirely 100% clear. A participant might have taken 9 years to go from university degree to a doctorate degree like a PhD, while another might have taken 20 years. We know the order of the two participants is the same (university degree to doctorate), but not the differences between the categories.

As previously mentioned, the first level of measurement are categorical variables, which are composed of categories or distinct things/entities and include nominal and ordinal variables. The second level of measurement is a step up in sophistication and is a continuous variable. A **continuous variable** is a variable that can take on any value within a range and can be split up into smaller units of measurement. These values can be whole or partial numbers (fractions). These variables are generally measured using a scale or measurement tool. A few common types of continuous variables you are familiar with include height, weight, time, temperature, and 5-star ratings. There are two kinds of continuous variables used in communication research: interval variables and ratio variables.

An **interval variable** has a clear and standardized distinction between values. Temperature is a classic example of an interval level variable. The difference in temperature between 10 and 15 degrees Celsius is the same difference in temperature as 15–20 degrees Celsius. In social sciences, Likert and Likert-type scales are classic examples of interval level variables. See Figure 2.3 for an illustration.

Strongly disagree	Disagree	Neutral	Agree	Strongly Agree
1	2	3	4	5

Figure 2.3 Sample Likert scale with interval items.

The difference between the points on this Likert scale are standardized, so that the difference between 1–2 and 3–4 is the same in practice, if not an exact measurement of distance. In addition, choice of a numeric value (let's say 5 – Strongly Agree) demonstrates two steps higher than 3 (Neutral), and this difference is meaningful to an observer. Most self-report and other-report data are based on interval level variables.

A **ratio variable** goes a step further than interval variables by requiring that there be a true zero point. The presence of a zero point on the scale makes it possible to compute/compare relationships in terms of proportions. An example of a ratio variable is height. Both Celsius and Fahrenheit temperature scales have a zero, but not a true zero. What a true zero means is an absolute lack of something (which for temperature is a theoretical construct called absolute zero on the Kelvin scale). Height is a ratio variable because it has a true zero. The value added of a true zero is that it allows you to make proportional comparisons. Because height has a true zero, we know that a person who is 3′ tall is exactly half the height of someone who is 6′ tall. If the common temperature scales were ratio data, then 60° F would feel exactly twice as warm as 30° F, but that is certainly not the case.

Students engage with ratio assessments daily. Any time you've taken an exam or quiz, you engaged in a form of cognitive measurement that produced ratio data. Because of this, the student who scores a 100 does twice as well on an exam as the student who scores a 50.

Test Your Understanding 2.4 – What are the differences between categorical and continuous variables?

Introduction to Reliability and Validity

Whenever you conduct research, you want to confidently say that your results are reliable and valid. Reliability and validity are key related concerns for researchers. Both terms have multiple meanings, and ways of determination. What follows below is a summary of both reliability and validity. In Chapters 17, and 18–20, you will dive deeper into each of these concepts, what they mean, and how to statistically measure them. For now, we will define reliability and validity and explain the different types of each in social scientific research.

Reliability is the extent to which a research tool, measure/instrument, survey, or process produces consistent results when applied repeatedly under similar conditions. Scales that measure weight are a good example of a "reliable" measure. If you stand on one, read your weight, get off, and a second later get back on and read your weight again, get off again, and repeat the process a few times, your weight should not change, unless you change clothes or drink a jug of water in between weighing. The scale will give the same weight each time over the few seconds, as it is a "reliable" instrument, unless you change. An unreliable scale would be one that within two seconds lists your weight as two different weights, and you have done nothing differently.

In social scientific research, which most of you reading this book will be engaged in, we often use survey-type measures to conduct our research. When using survey-type measures, we must ensure our assessments reliably measure the constructs we are studying. For example, imagine you are exploring organizational dissent in the workplace, and you are using the 18-item Organizational Dissent Scale by Kassing (2000). How do you know your participants read the 18 items and answered the questions with effort and care? Basically, how do you know the participants were consistent in their approach to answering the scale and that the answers are reliable? One of the first things you will need to do as a social scientific researcher is determine if your measure(s) is reliable. In Chapter 17, we will define and discuss different ways to test a measure's reliability: test-retest, alternate forms, split-half, Cronbach's alpha, and McDonald's omega.

Validity is the extent to which a research tool, instrument, questionnaire, or process assesses what it is supposed to measure. For example, would you use a spoon to check blood pressure? No, a spoon is an eating utensil, not a medical device used to check blood pressure; if you want to check blood pressure, you would use a blood pressure reader/monitor. It's the same kind of thinking when it comes to our measures in social scientific research. If you want to measure patient self-advocacy, you will not expect to see an item on a measure asking participants about the amount of credibility they have for their teacher. Essentially, whenever you see a researcher discuss "validity" of a measure/assessment, they are talking about whether it measures what it says it measures. There are various kind of validity, and ways to statistically establish and test for validity. The different kinds of validity (outlined in Chapter 18) are face, construct, content, concurrent, predictive, and discriminant. Chapters 18–20 describe how to statistically test for each of these forms of validity.

Test Your Understanding 2.5 – What are the differences between reliability and validity?

Data Sampling

It is not possible to collect all available evidence on a given topic. For example, let's say you are interested in exploring why New Zealanders blurt out things in a conversation or argument. Surveying every adult in New Zealand about their inclination to blurt and how it is perceived is impossible. Adults in New Zealand is your **population** or the group of units/cases from which you want to collect data. What you must do is collect a **sample** of adults in New Zealand, a representative group from the population. So, what you could do instead of trying to survey every adult in New Zealand, is give the 10-item Blurting Scale (Hample et al., 2013) to a sample of 200 adults in New Zealand. We would then base any of our conclusions off our sample. The purpose of sampling is to generate a sample that best represents the characteristics of the population, so you can make logical generalizations about the population. A **generalization** is an inference or judgment one makes about the population from analyzing the sample.

Generalizations from a sample to a population are possible because of something known as the **Central Limit Theorem**, which describes how, under normal conditions, the distribution of a sample mean converges to the mean of the population. There are three elements to the Central Limit Theorem to help unpack this definition. First, under normal conditions, data that come from larger samples tend to be more normally distributed. Second, when more samples are taken from the population, you increase the chances of accurately representing the population. Third, random selection is the most effective way for a sample to represent the population and reduce sampling error. Example 2.1 is an example of sampling from a population.

Example 2.1

A classic example of sampling from a population happens every four years in the United States on Election Night. Every four years, the media networks (CNN, FOX News, MSNBC, etc.) closely follow the elections in each U.S. state. As voters go to the polls to choose their preferred candidate for President, the networks track the incoming results. As the polls close (voting ends), networks track incoming results and "call" a state for a candidate. Often, states are called for a candidate with less than 10% of the votes counted. Your three authors have all been asked by students, "How is this possible?" What the networks and the political parties do is base their calling on a sampling of the population. None of the networks or political parties have the full voting results (population), but they have a large enough sample (they think) to make an inference (prediction) on who will win that state. So, based on the data they have and the trend they see, networks will announce the winner of a state before many of the votes are ever counted, which does not mean the result is official. This is not an exact science, though. In the 2000 Presidential Election, all the networks called the state of Florida for presidential candidate Al Gore, but the margin between him and George W. Bush was very slim. In the end, George W. Bush was officially declared the winner of the state. Recently, networks have been more careful about making their predictions and waiting for larger, more representative samples. For example, in the 2020 election, networks waited days to call states like Arizona, when normally all states would have been called on election night, as there was only a 0.4% difference between Joe Biden and Donald Trump.

When you are conducting social scientific research, as with any kind of research, your data sampling techniques should be organized. Once you have determined your research topic and population of study, you need to decide on how you will gather your sample. This can be done through a random or non-random sampling technique. A **random sample** is a sample in which everything or everyone that is part of your population has an equal chance of inclusion. A **non-random sample** is a sample in which not everyone or everything that is part of your population has an equal chance of participation. Random sampling is very hard to accomplish; in fact, the three of us would argue it is *almost* impossible in most situations. Figure 2.4 shows the advantages and disadvantages of each sampling technique. There are three kinds of random sampling techniques that researchers employ: simple, systematic, and stratified. **Simple random** sampling is a method in which everyone or everything from a population are selected entirely by chance, with each selection being independent of the others. This means that every member of the population has an equal probability of being chosen, and no external factors influence the selection process. Let's say you are interested in studying vaccine hesitancy among registered voters in a California county. A county office will have the name and address of all registered voters. Thus, you could mail a survey to all registered voters, and therefore, every registered voter would have an equal chance of participation. If your population was "potential" voters, this would not work, as you cannot guess who potential voters are. A simple random sample from that list would be one in which a computer system randomly chooses some percentage of the registered voters to mail (or you draw a percentage of voter names out of a giant bowl). However, there is a registry of registered voters, and thus, you can reach out to those on a list. Going off that same list of registered voters, let's say you choose to mail the survey to every nth person in your dataset; this is **systematic random**. In a **stratified random** sample, you identify mutually exclusive categories/groups and divide your sample into those categories/groups to select cases. Returning to our registered voters, you may choose to survey a sample of registered Republicans and registered Democrats, but not to survey registered Independents. Here is an example of a random sampling dilemma.

Example 2.2

Your communication department wants to measure the communication apprehension of all first-year communication majors. The department will have names and access information of all first-year communication majors. Someone in the department, after receiving proper ethical approval, can send each first-year communication major an email link to an apprehension scale. In this case, the population is small and easily manageable. If the population gets larger, such as New Yorkers, how can the researcher honestly say that *every* New Yorker has a chance to be contacted? Thus, simple random sampling with large populations, we assert, is virtually impossible.

With random sampling being virtually impossible, researchers more often use non-random sampling. There are four kinds of non-random sampling techniques: convenience, snowball, purposive, and quota. A **convenience sample** is when your data is easily accessible. For example, there have been numerous studies in the social sciences (including communication) where undergraduate students have received extra credit for completing surveys for their instructors. It's common to see statements such as: "224 undergraduate students from a Midwestern/Southern university completed a series of questions on X." Likely, the research participants came from the researcher's own department in such studies. Convenience sampling is the most used non-random sampling technique in the social sciences. Like convenience samples, **snowball sampling** is when a researcher contacts one participant, and from there, the sample builds off contacts from the initial participant.

Sampling Technique	Advantages	Disadvantages
Simple Random	Most generalizable	Expensive, time consuming, hard to do without list of population
Systematic Random	Generalizable and normally less expensive than simple random	At what starting point do you begin? Expensive, time consuming, hard to do without list of population
Stratified Random	Generalizable and you include specific targeted groups	Must carefully define groups, expensive, time consuming, hard to do without list of population
Convenience	Cheapest and easiest way to collect data	Is the least representative of the population
Snowball	Can more easily reach members of groups who are not normally included on lists	Hard to determine if the sample represents the population. Lack of generalizability
Purposive	You can ensure group sizes are balanced	Hard to determine if the sample represents the population. This method is open to researcher bias and subjectivity
Quota	You can ensure the selection of appropriate number of subjects with appropriate traits/characteristics	How do we know the sample represents the population? Possible lack of generalizability.

*Figure adapted from Croucher and Cronn-Mills (2022).

Figure 2.4 Advantages and disadvantages of different sampling techniques.

While this method is more commonly used in other research approaches, such as interviews and focus group research, it is also effective in acquiring additional participants for survey research. **Purposive sampling** is when you focus your study on a specific group and exclude others. For example, Ramsey (2024) studied university students' perceptions of physicians' humor usage. In this study, he was only interested in university students, and thus, he purposively sampled university students. As well as being a purposive sample, his study was also a convenience sample, as "The participants were ($N = 237$) undergraduates at a mid-sized, Mid-Atlantic university" (p. 29). **Quota sampling** is when you, as the researcher, pre-determine categories/groups and how much data (participants) you want in each category/group. Going back to our registered voters, you might say you want an equal number of registered Republicans and registered Democrats.

In Chapter 5, we will further discuss issues related to how we are able to sample from a population and make inferences with data.

<u>Test Your Understanding 2.6</u> – What are the differences between a random and non-random sample?
<u>Test Your Understanding 2.7</u> – How does a sample differ from the population?

Summary

In Chapter 2, we presented some key terms linked with the basics of measurement. We first discussed the differences between methodology and method. We then delineated between conceptualization and operationalization. The chapter then described the different methods of measurement. We also described the different types of variables and the different levels of measurement. The chapter also defined reliability and validity, explaining the differences between the two terms. Finally, we discussed data sampling. It is our hope that this chapter has provided you with further building blocks to help you on your statistical journey. You are better prepared now to lift some rocks, do some flips, and face the dark side in a cave, or, as we would put it, . . . read Chapter 3 on how to use SPSS and Excel for statistical analysis.

References

Cheney, G. (1983). On the various and changing meanings of organizational membership: Field study of organizational identification. *Communication Monographs*, *50*(4), 342–362. https://doi.org/10.1080/03637758309390174

Croucher, S. M., & Cronn-Mills, D. (2022). *Research methods: A theoretical and practical approach* (3rd ed.). Routledge.

Croucher, S. M., DeMaris, A., Diers-Lawson, A. R., & Roper, S. (2017). Self-reporting and the argumentativeness scale: An empirical examination. *Argumentation*, *31*(1), 23–43. https://doi.org/10.1007/s10503-015-9385-z

Croucher, S. M., Kassing, J. W., & Diers-Lawson, A. (2013). Accuracy, coherence, and discrepancy in self and other reports: Moving toward an interactive perspective of organizational dissent. *Management Communication Quarterly*, *27*(3), 425–442. https://doi.org/10.1177/0893318913476142

Croucher, S. M., Long, B. L., Meredith, M. J., Oommen, D., & Steele, E. L. (2009). Factors predicting organizational identification with intercollegiate forensics teams. *Communication Education*, *58*(1), 74–91. https://doi.org/10.1080/03634520802450523

Croucher, S. M., Murray, N., Ashwell, D., Cullinane, J., & Nguyen, T. (2022). Predicting vaccine confidence: Patient self-advocacy and fear of COVID-19 in the United States. *Southern Communication Journal*, *87*(4), 348–360. https://doi.org/10.1080/1041794X.2022.2099569

Fall, L. T., Kelly, S., Macdonald, P., Primm, C., & Holmes, W. (2013). Intercultural communication apprehension and emotional intelligence in higher education: Preparing business students for career success. *Business Communication Quarterly*, *76*(4), 412–426. https://doi.org/10.1177/1080569913501861

Glaser, S. R., Zamanou, S., & Hacker, K. (1987). Measuring and interpreting organizational culture. *Management Communication Quarterly*, *1*(2), 173–198. https://doi.org/10.1177/0893318987001002003

Guerrero, L. K., Anderson, P. A., Jorgenson, P. F., Spitzberg, B. H., & Eloy, S. V. (1995). Coping with the green-eyed monster: Conceptualizing and measuring communicative responses to romantic relationships. *Western Journal of Communication*, *59*(4), 270–304. https://doi.org/10.1080/10570319509374523

Hample, D., Richards, A. S., & Skubisz, S. (2013). Blurting. *Communication Monographs*, *80*(4), 503–532. https://doi.org/10.1080/03637751.2013.830316

Kassing, J. W. (2000). Investigating the relationship between superior-subordinate relationship quality and employee dissent. *Communication Research Reports*, *17*(1), 58–70. https://doi.org/10.1080/08824090009388751

Kelly, S., & Westerman, D. K. (2024). *A primer in the philosophy of science: A guide to thinking like a scientist for social scientists*. Kendall Hunt.

Kim, L. T., & Oh, S.-H. (2023). Predicting telephone anxiety: Use of digital communication technologies, language and cultural barriers, and preference for phone calls. *Communication Research Reports*, *40*(3), 156–168. https://doi.org/10.1080/08824096.2023.2221845

McCroskey, J. C., & Teven, J. J. (1999). Goodwill: A reexamination of the construct and its measurement. *Communication Monographs*, *66*(1), 90–103. https://doi.org/10.1080/03637759909376464

Neuliep, J. W., & McCroskey, J. C. (1997). The development of intercultural and interethnic communication apprehension scales. *Communication Research Reports*, *14*(2), 145–156. https://doi.org/10.1080/08824099709388656

Neuman, W. L. (2011). *Social research methods: Qualitative and quantitative approaches* (7th ed.). Routledge.

Nicotera, A. M. (1996). An assessment of the argumentativeness scale for social desirability bias. *Communication Reports*, *9*(1), 23–25. https://doi.org/10.1080/08934219609367632

Pew Research Center. (2021, May 14). www.pewresearch.org/social-trends/2021/05/14/only-about-half-of-americans-say-census-questions-reflect-their-identity-very-well/

Philipsen, G. (1976). Places for speaking in Teamsterville. *Quarterly Journal of Speech*, *62*(1), 15–25. https://doi.org/10.1080/00335637609383314

Podsakoff, P. M., & Organ, D. W. (1986). Self-reports in organizational research: Problems and prospects. *Journal of Management*, *12*(4), 531–544. https://doi.org/10.1177%2F014920638601200407 8

Ramsey, M. C. (2024). An examination of university students' perceptions of physicians' humor usage, source credibility, and fear of physician. *Communication Research Reports*, *41*(1), 25–35. https://doi.org/10.1080/08824096.2023.2292671

Sadiq, M., Croucher, S. M., & Dutta, D. (2023). COVID-19 vaccine hesitancy: A content analysis of Nigerian YouTube videos. *Vaccines*, *11*(6), 1057. https://doi.org/10.3390/vaccines11061057

StatsNZ. (2020, September 3). Ethnic group summaries reveal New Zealand's multicultural make-up. www.stats.govt.nz/news/ethnic-group-summaries-reveal-new-zealands-multicultural-make-up

Chapter 3

SPSS and Excel

What Will I Learn in This Chapter?

Lego Tony Stark, known as Legoland's ultimate billionaire, genius, playboy, and philanthropist, has never needed Excel or SPSS. With his programming prowess, he can crunch data faster than those programs can start up. Yet, here he finds himself explaining these programs to his new intern, who was hired for his brilliance and curiosity, but not his technical skills. As Lego Tony Stark rubs his temples, he explains, "Okay, Kid, think of Excel as a flashy sports car. It's got all the bells and whistles, but sometimes you need more horsepower to get a job done. SPSS is a tank. It's big and clunky, but far more practical and reliable than that sports car when you need to do heavy lifting. Bottom line, know your tools. These are going to be the best programs for you to start with because they let you do a lot of the work by pointing and clicking your mouse rather than learning a computer language.

DOI: 10.4324/9781003500117-3

You'll want to use Excel for quick tasks, but SPSS for the more heavy-duty analyses. Then, when you're ready for your data power suit, you'll learn how to code." In this chapter, you will learn:

1. When you would choose to use SPSS versus Excel
2. How to get started using Excel
3. How to get started using SPSS

Excel versus SPSS

Excel is Microsoft's general-purpose spreadsheet. You can use it for everything from making your grocery list to basic statistical analyses. In addition to being a multi-purpose spreadsheet, Excel is also designed for people who are not data savvy. It offers useful tips and help as you work.

In terms of data analysis, Excel is the software you want to use for any analysis that you would feel comfortable doing with the assistance of a calculator and a notebook. If you want basic descriptive statistics or charts, Excel is a wonderful tool. Excel is also an excellent place to sort and clean data.

The best thing about Excel is that you'll probably always have access to it. Many workplaces and all universities in the United States (and many overseas) have Microsoft 365 accounts, which means Excel comes standard issue. In a pinch, you can do some advanced analyses in Excel, but eventually, the difference between what Excel *can* do loses the battle with whether it's worth your time to *make* Excel do those things. Eventually, we'll hit a point in this book where we quit showing you how to do things in Excel because it would be less of a headache for you to use AI to write a brief statistical program to analyze your data in R or Python rather than making Excel try to do a heavy lift and give you output in a way you can easily understand.

SPSS is designed to help researchers. In fact, SPSS is an acronym for the *Statistical Package for Social Sciences*. Recall the studies you've read about so far in this book, such as Fall et al.'s (2013) research on ICA or Croucher et al.'s (2009) work on organizational identity. The data for these studies were cleaned and analyzed in SPSS. **Data cleaning** is the process of performing quality inspections on data (e.g., managing missing items, removing duplicates, correcting typos, formatting, removing data that fail attention checks). **Analyzing data** involves performing calculations with the data to answer questions.

SPSS is the tool you need to handle advanced statistical analyses, transform data, generate detailed statistical reports, conduct hypothesis testing, and conduct data modeling. SPSS can also do all the things Excel does, except for some more advanced chart builds. The drawbacks of SPSS are that it is not as user-friendly as Excel, and it is expensive. You'll likely always have access to it as a university student or university employee, but otherwise, you're not as likely to have access to it as you are to Excel. Nonetheless, you're a student now, so we're going to see this book all the way through with how to conduct analyses in SPSS.

Test Your Understanding 3.1 – When should you choose Excel and when should you choose SPSS?

Using Excel

When you open Excel, you are shown a general data management spreadsheet. You have rows and columns that you can add data to. Excel assumes that row 1 of your data is the name of the variables. You can break your data up into more variables by adding sub-headers or adding additional labels to the first column. For effective data analyses, though, the best setup is to have one variable label for each column

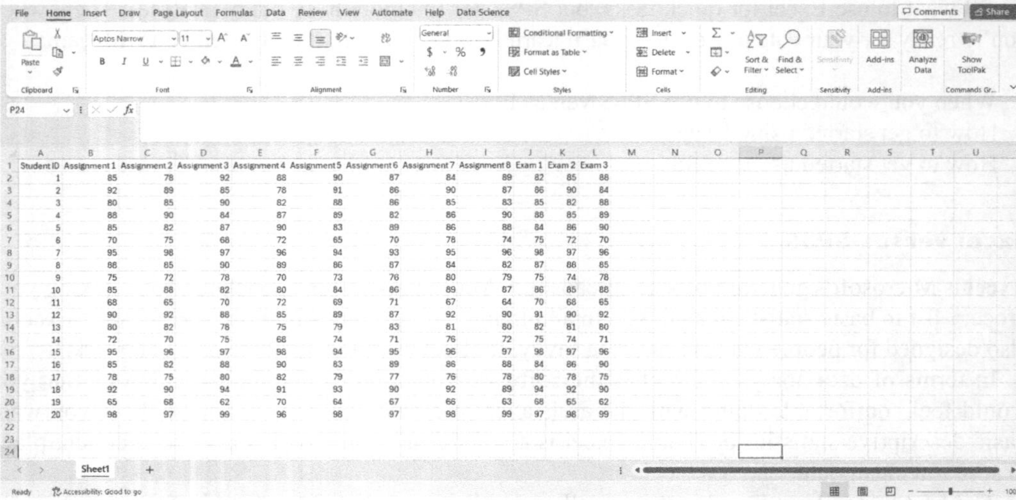

Figure 3.1 Basic Excel layout.

of data with no additional labels. Figure 3.1 shows the basic layout in Excel. In this figure, grades from a course with six assignments and three exams are shown. (This data is created, not real class data.) Notice that each column has a label telling you what data is contained in that column.

Before analyzing data in Excel, it is highly recommended that you format your data as a **table**. As far as Excel is concerned, a table is data organized into columns with a label for each column. The label in each column that tells users what type of data appears in the column is called a **header**. To format Excel data as a table, follow the steps below and use Figure 3.2 as a visual guide:

1. Left click on the data. It does not matter where you click; if you click somewhere, you've included information.
2. Select all data. Do this on a PC or Chromebook by clicking CTRL + A or on a Mac by clicking Command + A.
3. Click Format as a Table.
4. Select a Table design you like.
5. Agree with Excel when it says it looks like you have headers in your data by clicking OK.

Once your data is formatted as a table, you have many options for sorting. Each header should now have a dropdown arrow in it (see Figure 3.3). The dropdown arrow allows you to sort and filter the table. The three things you are likely to use are as follows:

1. Sort Largest to Smallest: This option will organize all your data in descending order of the column you chose.
2. Sort Smallest to Largest: This option will organize all your data in ascending order of the column you chose.
3. The check boxes: Here, any boxes that are checked mean that rows with these response options will be displayed in your data. Imagine that I only want to look at data for students who earned below a C on the first assignment. I can uncheck any grades C or higher here, and that will leave only data for those students who earned below a C.

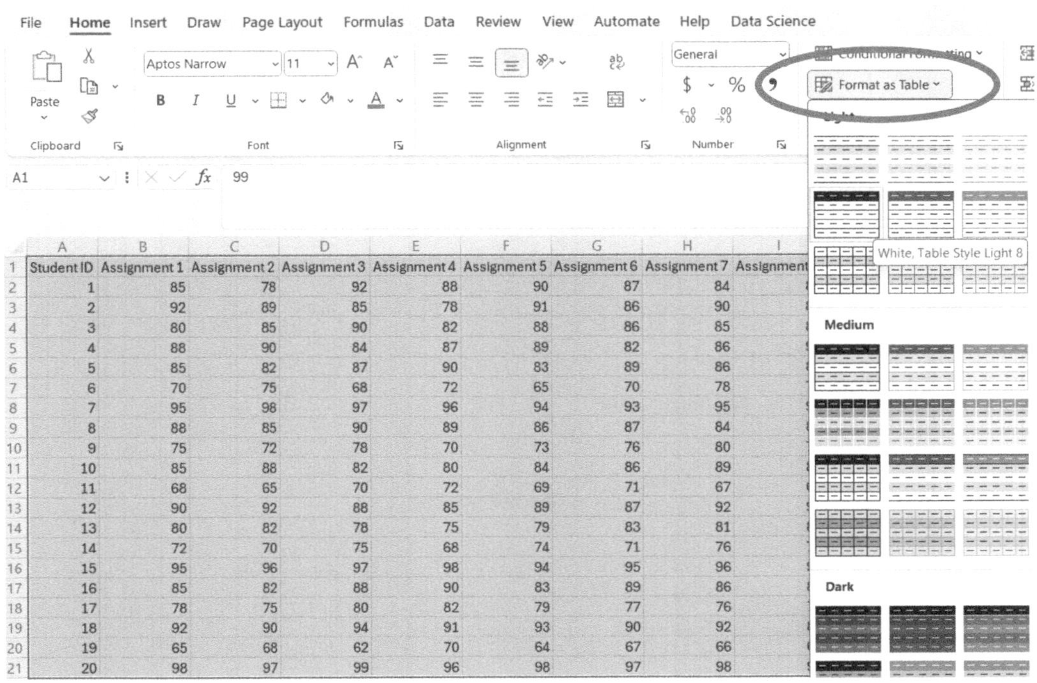

Figure 3.2 Formatting as a table.

These sorting and filtering options often come in handy!

Excel will also allow you to conduct analyses by writing formulas. If you've used a graphing calculator, you can treat Excel much like your graphing calculator. One small exception to this is that in Excel, you must begin every formula by typing an equal sign. In cell M2 of Figure 3.4, the average of exam scores is calculated. This can be done by a manual formula, adding the scores for all three exams and dividing by the number of exams:

$$= (82 + 85 + 88)/3$$

Excel does quick arithmetic, so typing in the literal math will always work. You can also have Excel find the numbers for you by typing in the cell names these numbers are in instead. In the data, the exam scores are listed in columns J, K, and L, with the first student's scores in row 2. This means the exam data for that first student is in cells J2, K2, and L2. Using the cell names, the formula looks like this:

$$= (J2 + K2 + L2)/3$$

Using cell names is the recommended method of performing arithmetic tasks in Excel for several reasons. First, if you identify the cells, your average cell will update automatically if you find later that there was a typo in your data. Second, as you type the cell names, they will change colors so that you can be sure you are selecting the exact right cells. This is a way in which Excel is user-friendly and helps prevent you from making typos. Third, Excel provides a

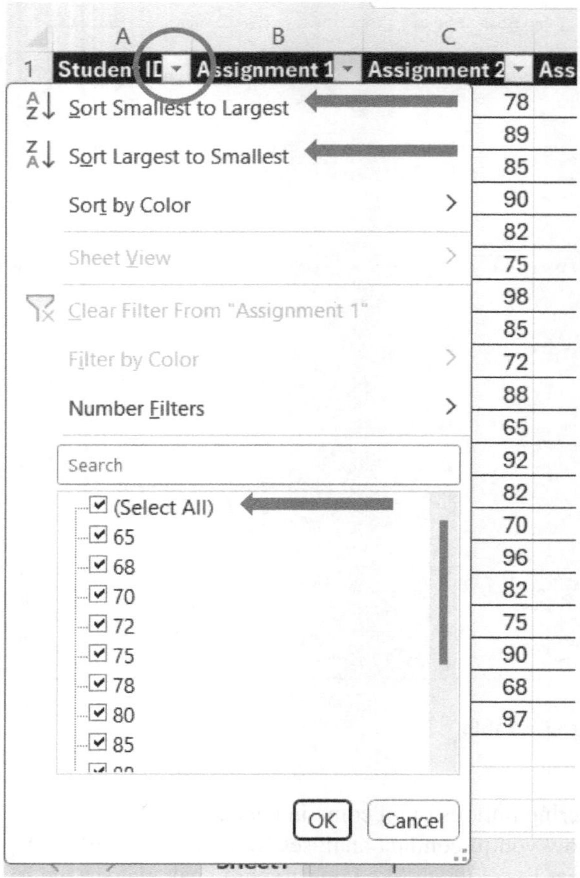

Figure 3.3 Using a table.

shortcut for you when you use this method. Figure 3.4 shows what your data will look like once you've typed in this formula and clicked Enter. Notice the little square that appears at the bottom of M2. Double-clicking that little box fills the formula in for the rest of the column without you needing to do any more typing (i.e., the next row is automatically filled with =(J3+K3+L3)/3, and so forth).

If you want to be fancy, instead of typing in the names of your cells, you can click on the cells one at a time to add them to your formula. That's a bit hard to show you how to do in a book, though, so let's stick with the method of typing our cell names for these examples.

Another lovely thing about Excel is the Formula ribbon. Excel has several built-in formulas that give you even more shortcuts. Using the Insert Function or AutoSum tools will allow you to find all the built-in functions Excel has for you. (A **function** is a formula that Excel has saved in shortcut form. Once you know the names of these functions, you can also just type them in instead of an equation!) The function Excel has for the arithmetic we've been doing is called, no surprise, AVERAGE. In Figure 3.5, the AVERAGE function is used in N2 to create the average assignment

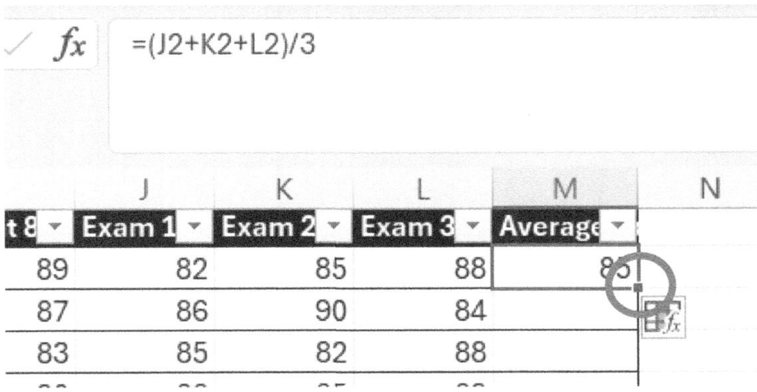

Figure 3.4 Filling a formula.

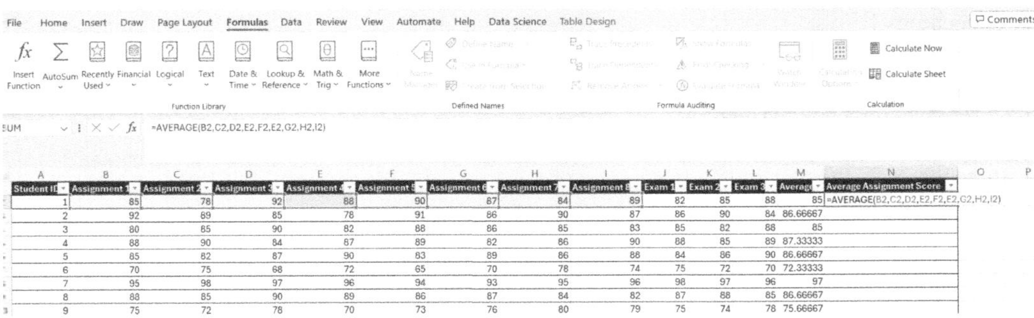

Figure 3.5 Formula tab.

score. This means that instead of typing out the math, the function name is typed, followed by the location of each piece of data, separated by a comma:

= AVERAGE(B2,C2,D2,E2,F2,E2,G2,H2,I2)

You can use the fill tool using functions just like when you type out the formula yourself. These types of simple data manipulations are ideal for using Excel!

Test Your Understanding 3.2 – Why is it useful to format your data as a table when using Excel?

Data Analysis Menu

Most of your work in Excel will require using the Data Analysis menu. This is a pop-up menu that you find using the Data tab. You will need to find this menu many times throughout this class if you use Excel for your work, so take note. The Data Analysis menu is found as shown in Figure 3.6.

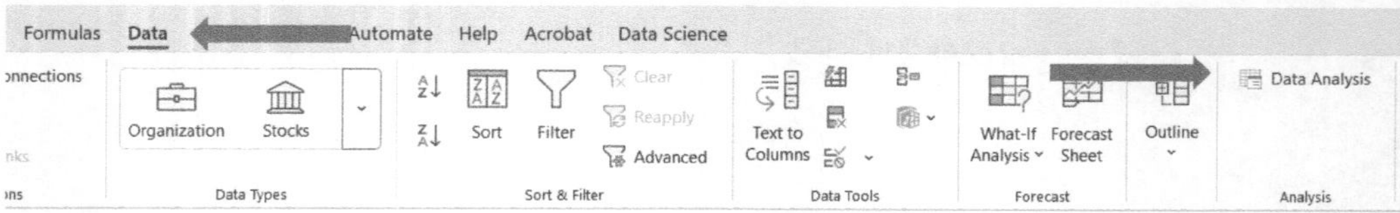

Figure 3.6 Excel Data Analysis menu.

Reverse Coding

Later in this book, you'll discuss reverse coding measure items. For now, suffice it to say that this process requires flipping the values of certain responses. Imagine you've used a measure where respondents indicate their agreement on a 7-point scale: 1 = Disagree Strongly, 2 = Disagree, 3 = Disagree Somewhat, 4 = Neutral, 5 = Agree Somewhat, 6 = Agree, and 7 = Agree Strongly. If you need to reverse-code the items, that means:

- 1 becomes 7
- 2 becomes 6
- 3 becomes 5
- 4 remains unchanged (neutral)
- 5 becomes 3
- 6 becomes 2
- 7 becomes 1

To perform this transformation in Excel, you can use an IF statement in a blank column. This formula instructs Excel on how to modify each value when it encounters a response that needs to be changed. The example below assumes the data to be reverse-coded is in column B. If your data is in a different column, you'll need to adjust the formula accordingly.

= IF(B2 = 7,1, IF(B2 = 6,2, IF(B2 = 5,3, IF(B2 = 4,4, IF(B2 = 3,5, IF(B2 = 2,6, IF(B2 = 1,7,”“)))))))

Using SPSS

When you first open SPSS, you're going to be hit by a lot of pop-up windows asking you where you want to start. Feel free to exit out of all of them. If you have an existing dataset, you can click File -> Open just like any other program to open your existing file. If you do not have a new dataset, those pop-ups are just in your way. When you get the pop-ups closed, you'll be in a spreadsheet that will remind you a lot of Excel. In the lower left corner, you'll see two or three tabs (depending on which version of SPSS you're using; see Figure 3.7). The two tabs you need to become familiar with, no matter which version of SPSS you're using, are Data View and Variable View.

The spreadsheet part of SPSS is called the **Data View**. Here, you can enter data one cell at a time, copy and paste data from a spreadsheet like Excel, or import data from a spreadsheet. To get started analyzing the same grade data in SPSS, we can click on File -> Import Data -> Excel. Because the data from Excel is already formatted as a table, SPSS knows just what to do with it, and you only need to click "OK" to get the data read in. If the data were not formatted as a table, you would copy and paste the data into SPSS because otherwise, the transfer could be a disaster.

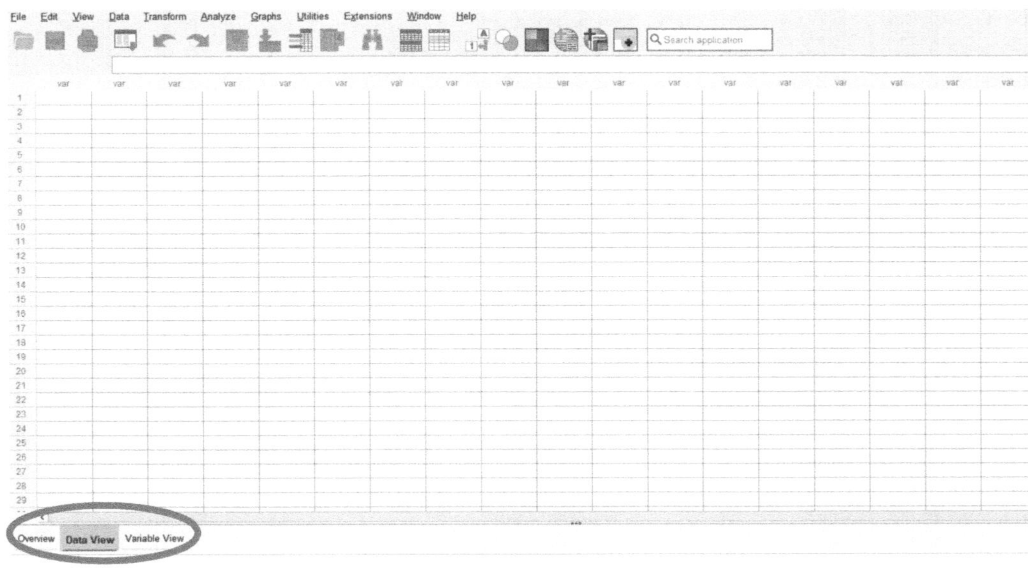

Figure 3.7 Opening SPSS.

Table 3.1 Student Gender

Student 1	1
Student 2	1
Student 3	2
Student 4	1
Student 5	2
Student 6	1
Student 7	1
Student 8	1
Student 9	1
Student 10	1
Student 11	1
Student 12	1
Student 13	3
Student 14	1
Student 15	1
Student 16	2
Student 17	2
Student 18	2
Student 19	2
Student 20	2

Imagine that in addition to the grade data, you also have gender data on our students. The gender data is shown in Table 3.1. Here, student gender is coded such that 1 = male, 2 = female, and 3 = third gender. Here is a place where Excel is more user-friendly than SPSS. We must code all of our qualitative (word) data so that it goes into SPSS as numbers. SPSS does not speak English (or any other language besides numbers). We can copy and paste the second column of that data into our data imported from

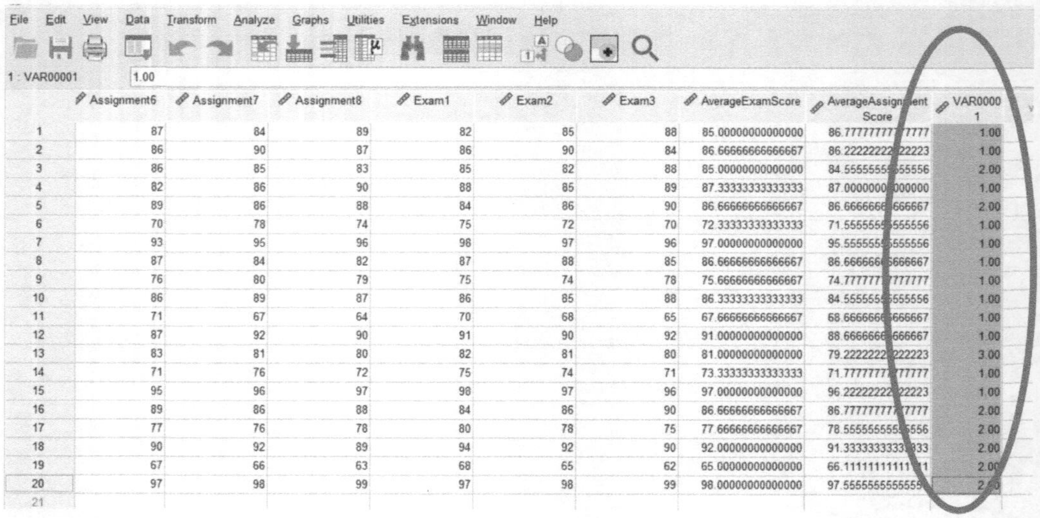

	Assignment6	Assignment7	Assignment8	Exam1	Exam2	Exam3	AverageExamScore	AverageAssignment Score	VAR00001
1	87	84	89	82	85	88	85.00000000000000	86.77777777[illegible]7777	1.00
2	86	90	87	86	90	84	86.66666666666667	86.22222222[illegible]22223	1.00
3	86	85	83	85	82	88	85.00000000000000	84.5555555[illegible]55556	2.00
4	82	86	90	88	85	89	87.33333333333333	87.0000000[illegible]000000	1.00
5	89	86	88	84	86	90	86.66666666666667	86.6666666[illegible]666667	2.00
6	70	78	74	75	72	70	72.33333333333333	71.5555555[illegible]555556	1.00
7	93	95	96	98	97	96	97.00000000000000	95.5555555[illegible]555556	1.00
8	87	84	82	87	88	85	86.66666666666667	86.6666666[illegible]666667	1.00
9	76	80	79	75	74	78	75.66666666666667	74.777777[illegible]777777	1.00
10	86	89	87	86	85	88	86.33333333333333	84.5555555[illegible]555556	1.00
11	71	67	64	70	68	65	67.66666666666667	68.6666666[illegible]666667	1.00
12	87	92	90	91	90	92	91.00000000000000	88.6666666[illegible]666667	1.00
13	83	81	80	82	81	80	81.00000000000000	79.2222222[illegible]222223	3.00
14	71	76	72	75	74	71	73.33333333333333	71.7777777[illegible]77777	1.00
15	95	96	97	98	97	96	97.00000000000000	96.22222222[illegible]22223	1.00
16	89	86	88	84	86	90	86.66666666666667	86.77777777[illegible]7777	2.00
17	77	76	78	80	78	75	77.66666666666667	78.5555555555[illegible]556	2.00
18	90	92	89	94	92	90	92.00000000000000	91.3333333333[illegible]33	2.00
19	67	66	63	68	65	62	65.00000000000000	66.11111111111[illegible]1	2.00
20	97	98	99	97	98	99	98.00000000000000	97.5555555555555[illegible]	2.[illegible]
21									

Figure 3.8 Loading data in SPSS.

Excel to get the spreadsheet shown in Figure 3.8. (Before merging datasets, be sure that each dataset is in the same order. If you did not know with certainty that the gradebook data was organized in the same way that Table 3.1 is, then you could be assigning the wrong gender code to students.)

Notice that because this data was not imported from an Excel table, SPSS has no idea what the numbers in the new column represent. To tell SPSS what this new information means, you must use the other tab of interest: **Variable View**. Variable View serves two purposes. It is where you make the spreadsheet more aesthetically appealing and where you can explain to SPSS what it has.

Figure 3.9 shows the Variable View in SPSS. There are several important columns to pay attention to here.

1. Name: This is the name of your variable as SPSS understands it. A limitation of SPSS is that it breaks down when it finds a space in information. Notice that when SPSS imported data from Excel, it took the spaces out of all the variable names. The new variable is Gender. In row 17 of Variable View under Name, you can delete the default variable name and replace it with whatever you want (as long as it has no spaces). In Figure 3.9, "Gender" should replace the default variable name by clicking in the cell and typing over it.
2. Type: The second important column is Type. Generally, you'll have Numeric data in SPSS. It may be that you have String data (i.e., words) at times, though. When this happens, know that you will need to recode your data to numbers if you want SPSS to analyze it well. (You will see how to do that soon!)
3. Width: This tells you literally how many characters can be typed in that column in the Data View.
4. Decimals: This option dictates how many decimals are shown with each variable. For this example, the cells with the Assignment and Exam Averages have been clicked on to reduce them to one decimal. When you click on a number in this column, SPSS has an up/down arrow pop-up that lets you change the decimal number.

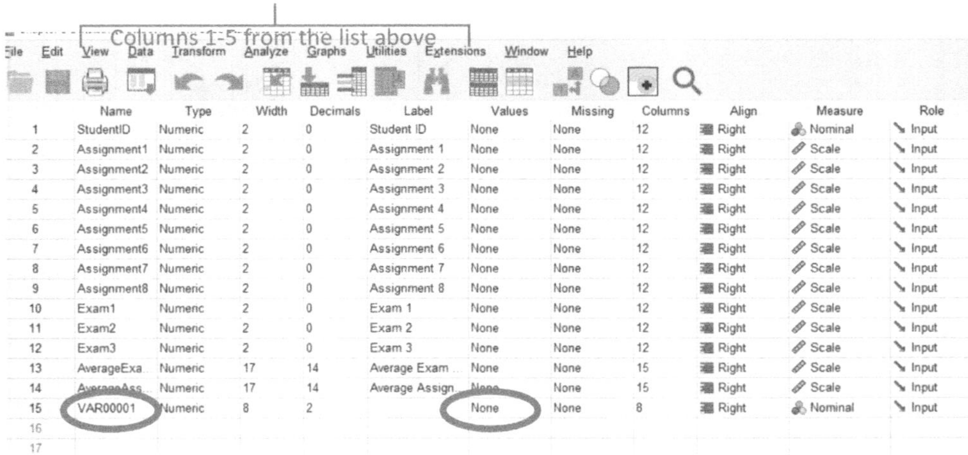

	Name	Type	Width	Decimals	Label	Values	Missing	Columns	Align	Measure	Role
1	StudentID	Numeric	2	0	Student ID	None	None	12	Right	Nominal	Input
2	Assignment1	Numeric	2	0	Assignment 1	None	None	12	Right	Scale	Input
3	Assignment2	Numeric	2	0	Assignment 2	None	None	12	Right	Scale	Input
4	Assignment3	Numeric	2	0	Assignment 3	None	None	12	Right	Scale	Input
5	Assignment4	Numeric	2	0	Assignment 4	None	None	12	Right	Scale	Input
6	Assignment5	Numeric	2	0	Assignment 5	None	None	12	Right	Scale	Input
7	Assignment6	Numeric	2	0	Assignment 6	None	None	12	Right	Scale	Input
8	Assignment7	Numeric	2	0	Assignment 7	None	None	12	Right	Scale	Input
9	Assignment8	Numeric	2	0	Assignment 8	None	None	12	Right	Scale	Input
10	Exam1	Numeric	2	0	Exam 1	None	None	12	Right	Scale	Input
11	Exam2	Numeric	2	0	Exam 2	None	None	12	Right	Scale	Input
12	Exam3	Numeric	2	0	Exam 3	None	None	12	Right	Scale	Input
13	AverageExa...	Numeric	17	14	Average Exam ...	None	None	15	Right	Scale	Input
14	AverageAss...	Numeric	17	14	Average Assign...	None	None	15	Right	Scale	Input
15	VAR00001	Numeric	8	2		None	None	8	Right	Nominal	Input
16											
17											

Figure 3.9 Variable View.

5. Label: The label is what SPSS knows you call the variables. Here, you can have spaces. When SPSS analyzes your data, it will present it to you based on what you have typed in Label.
6. Values: Values is one of the most important columns. Here, you will tell SPSS what your data coding means for any of your variables that are words that have been coded as numbers. If you click on the lower-right corner of the Value cell for gender, the window shown in Figure 3.10 pops up. Once this window pops up, you will click the plus sign to add a value to the data. For each value (i.e., each number code option in the data), type the number in the Value column and what the number corresponds to in the Label column. Because we have three gender options in this data, we have three labels. Once all labels are input, click OK.
7. Measure: Each of your variables will have a measure type assigned to it. If you hover your mouse over the measure type for any variable, you will receive a dropdown menu with the options: Nominal, Ordinal, or Scale. This is shown in Figure 3.11. Any categorical variable without rank should be set to Nominal. Any categorical variable with rank (e.g., first year, sophomore, junior, senior) should be set to Ordinal. All your numeric data that do not represent categories should be set to Scale.

Figure 3.12 shows how the Variable View will look for your data after these setup procedures in the Variable View, and Figure 3.13 shows how the Data View looks after these procedures.

Just like in Excel, SPSS will also allow you to create new columns of data based on your existing columns. Imagine that final grades in this sample class data are calculated with Assignments counting for 50% of the final grade and Exams counting for 50% of the final grade. You can create a new column that shows the final course grade for students by going to Transform -> Compute Variable (see Figure 3.14).

This will open a new window, shown in Figure 3.14, where you can use formulas like you did in individual cells of Excel. Under Target Variable, give your variable a name (but don't use spaces!).

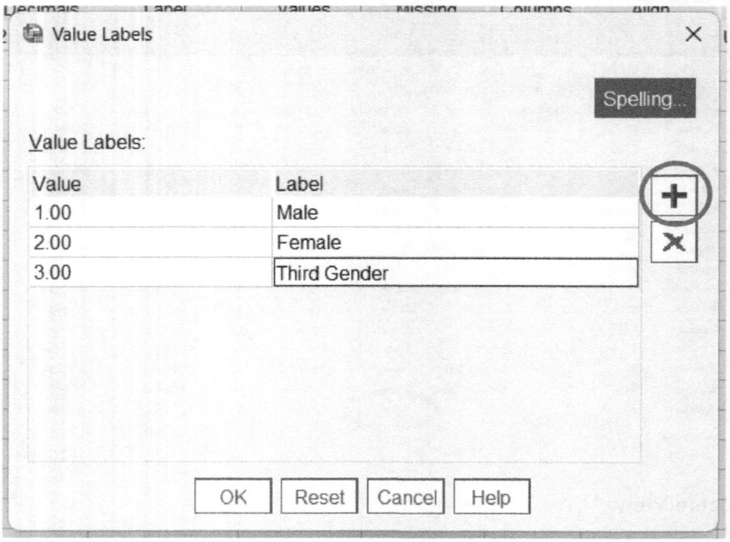

Figure 3.10 Values.

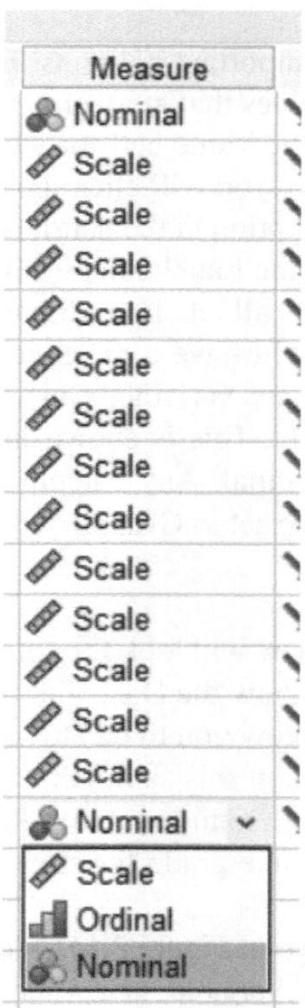

Figure 3.11 Measures.

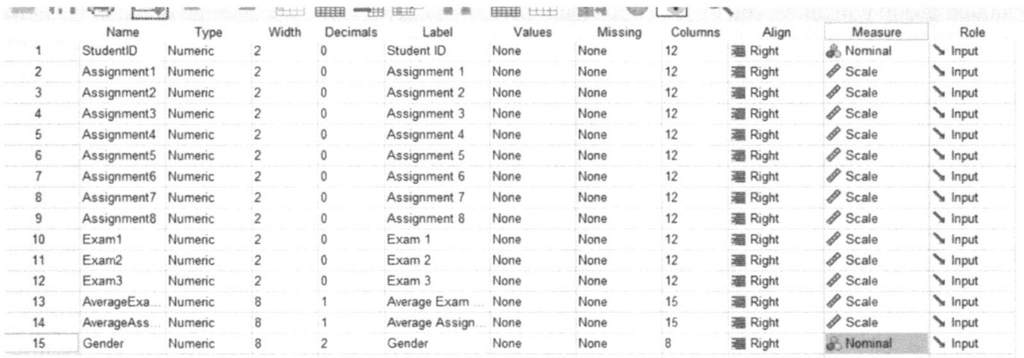

	Name	Type	Width	Decimals	Label	Values	Missing	Columns	Align	Measure	Role
1	StudentID	Numeric	2	0	Student ID	None	None	12	Right	Nominal	Input
2	Assignment1	Numeric	2	0	Assignment 1	None	None	12	Right	Scale	Input
3	Assignment2	Numeric	2	0	Assignment 2	None	None	12	Right	Scale	Input
4	Assignment3	Numeric	2	0	Assignment 3	None	None	12	Right	Scale	Input
5	Assignment4	Numeric	2	0	Assignment 4	None	None	12	Right	Scale	Input
6	Assignment5	Numeric	2	0	Assignment 5	None	None	12	Right	Scale	Input
7	Assignment6	Numeric	2	0	Assignment 6	None	None	12	Right	Scale	Input
8	Assignment7	Numeric	2	0	Assignment 7	None	None	12	Right	Scale	Input
9	Assignment8	Numeric	2	0	Assignment 8	None	None	12	Right	Scale	Input
10	Exam1	Numeric	2	0	Exam 1	None	None	12	Right	Scale	Input
11	Exam2	Numeric	2	0	Exam 2	None	None	12	Right	Scale	Input
12	Exam3	Numeric	2	0	Exam 3	None	None	12	Right	Scale	Input
13	AverageExa...	Numeric	8	1	Average Exam ...	None	None	15	Right	Scale	Input
14	AverageAss...	Numeric	8	1	Average Assign...	None	None	15	Right	Scale	Input
15	Gender	Numeric	8	2	Gender	None	None	8	Right	Nominal	Input

Figure 3.12 Updated Data View.

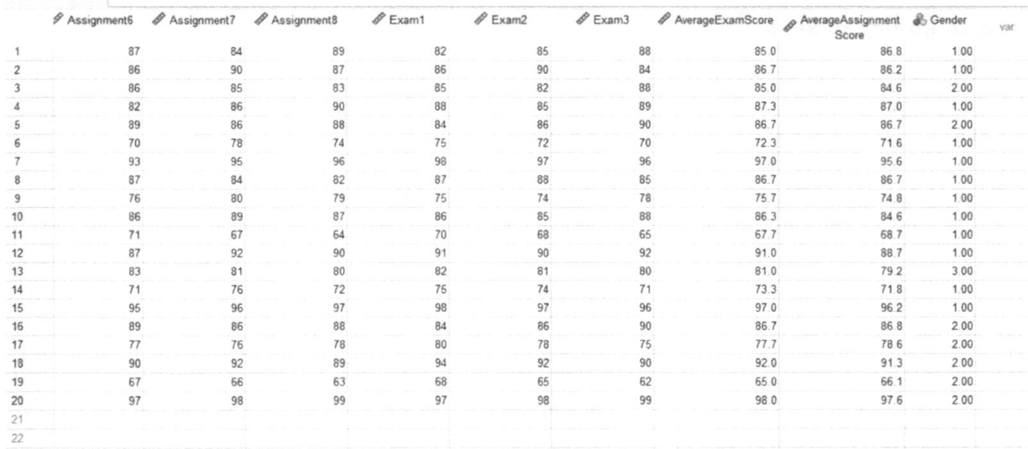

	Assignment6	Assignment7	Assignment8	Exam1	Exam2	Exam3	AverageExamScore	AverageAssignment Score	Gender	var
1	87	84	89	82	85	88	85.0	86.8	1.00	
2	86	90	87	86	90	84	86.7	86.2	1.00	
3	86	85	83	85	82	88	85.0	84.6	2.00	
4	82	86	90	88	85	89	87.3	87.0	1.00	
5	89	86	88	84	86	90	86.7	86.7	2.00	
6	70	78	74	75	72	70	72.3	71.6	1.00	
7	93	95	96	98	97	96	97.0	95.6	1.00	
8	87	84	82	87	88	85	86.7	86.7	1.00	
9	76	80	79	75	74	78	75.7	74.8	1.00	
10	86	89	87	86	85	88	86.3	84.6	1.00	
11	71	67	64	70	68	65	67.7	68.7	1.00	
12	87	92	90	91	90	92	91.0	88.7	1.00	
13	83	81	80	82	81	80	81.0	79.2	3.00	
14	71	76	72	75	74	71	73.3	71.8	1.00	
15	95	96	97	98	97	96	97.0	96.2	1.00	
16	89	86	88	84	86	90	86.7	86.8	2.00	
17	77	76	78	80	78	75	77.7	78.6	2.00	
18	90	92	89	94	92	90	92.0	91.3	2.00	
19	67	66	63	68	65	62	65.0	66.1	2.00	
20	97	98	99	97	98	99	98.0	97.6	2.00	
21										
22										

Figure 3.13 Updated Variable View.

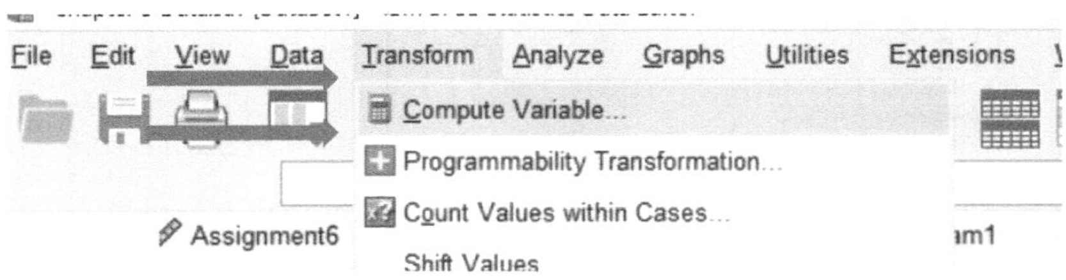

Figure 3.14 Compute Variable.

Function group will allow you to search for the function you need. Here, the average function is named *MEAN*, so you can calculate the average of those grade types by typing in the Numeric Expression box:

MEAN(AverageExamScore,AverageAssignmentScore)

Calculating the variable this way is shown in Figure 3.15. Alternatively, you can also use mathematical expressions just like in Excel. In Numeric Expression, you could type:

(AverageExamScore+AverageAssignmentScore)/2

Just be careful not to use spaces. On that note, if you ever get an error message from SPSS that says you've used an "illegal character," it's telling you that you have a space somewhere or are missing a comma.

Test Your Understanding 3.3 – When do you need to be in the Data View of SPSS, and when do you need to be in the Variable View?

Reverse Coding in SPSS

In SPSS, you can recode variables easily. In fact, SPSS assumes you may need to do this and provides two options for managing it. Under the Transform menu, you can choose Recode into Same Variables. . . or Recode into Different Variables. . .

- Recode into Same Variables. . . modifies the original column(s) of data.
- Recode into Different Variables. . . creates new columns for each reverse-coded item.

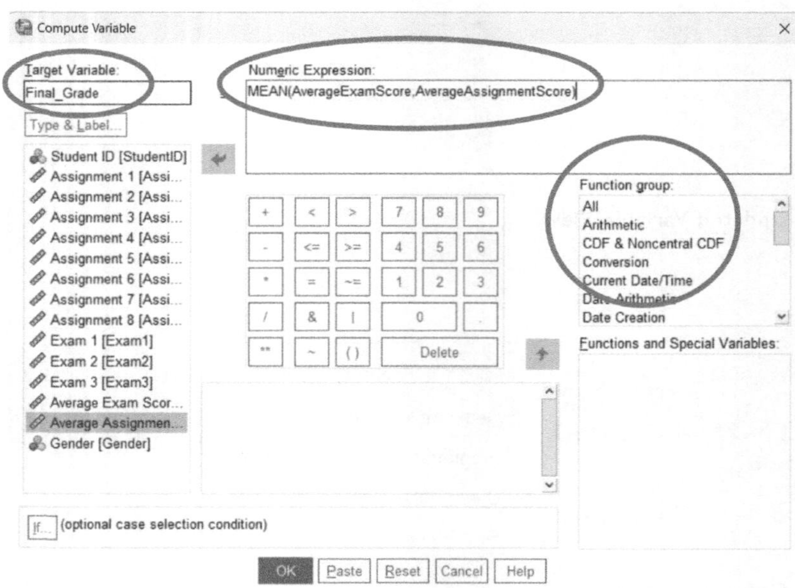

Figure 3.15 Final Grade.

As shown in Figure 3.16, these options allow you to either overwrite your existing data or create new columns, depending on your needs.
Regardless of which option you choose, the steps for reverse coding are the same:

1. Select the variables to be reverse-coded and move them into the Numeric Variables box
2. Click Old and New Values. . . (shown in Figure 3.17)

Once Old and New Values. . . is selected, the menu shown in Figure 3.18 will appear. In this window, you'll specify how SPSS should recode the values. While you may use different old value options in your research, the most common for reverse coding is *Value*, which allows you to enter specific values manually.

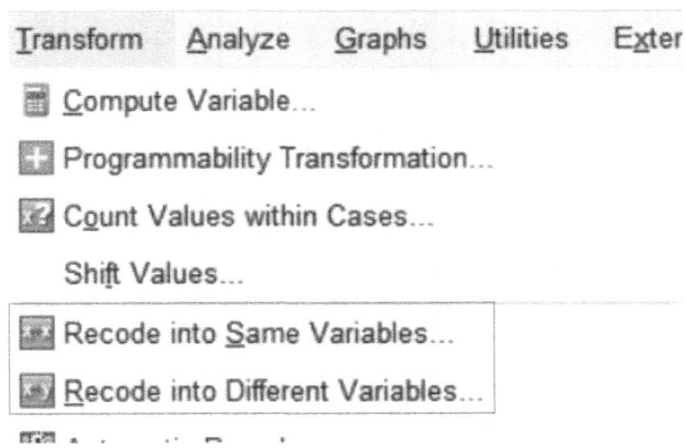

Figure 3.16 Reverse coding options.

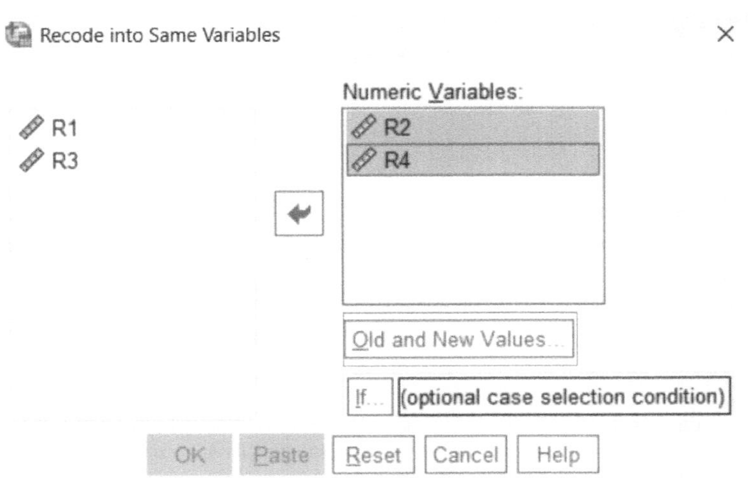

Figure 3.17 Reverse code steps.

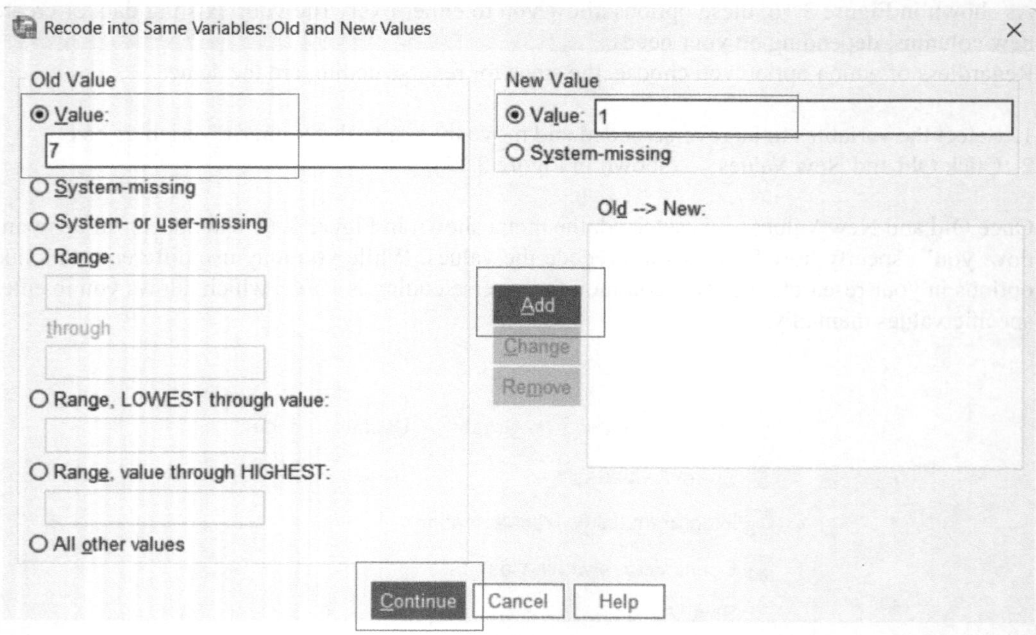

Figure 3.18 Old and New Values.

To complete the process:

3. Type one of your values into the Value field under Old Value
4. Enter its corresponding reverse-coded value in the Value field under New Value
5. Click Add
6. Repeat this process until all necessary value changes have been entered
7. Click Continue
8. Click OK

Summary

Excel and SPSS are probably the most user-friendly tools for beginning to learn data analysis. They both require great specificity with what you input, though. Like Tony Stark's new intern, you will probably run into a lot of frustrations when you first start using these programs, as you try to remember where all the tools live and encounter strange error messages. You will find that the more practice you have, the easier the programs become!

References

Croucher, S. M., Long, B. L., Meredith, M. J., Oommen, D., & Steele, E. L. (2009). Factors predicting organizational identification with intercollegiate forensics teams. *Communication Education*, *58*(1), 74–91. https://doi.org/10.1080/03634520802450523

Fall, L. T., Kelly, S., Macdonald, P., Primm, C., & Holmes, W. (2013). Intercultural communication apprehension and emotional intelligence in higher education: Preparing business students for career success. *Business Communication Quarterly*, *76*(4), 412–426. https://doi.org/10.1177/1080569913501861

Chapter 4

Data Visualization

What Will I Learn in This Chapter?

In the heart of Lego Land stands the Sanctum Sanctorum. Lego Dr. Strange calls this place his home but also his charge. In this sacred place, he has called Lego Thor, Lego Captain America, Lego Scarlet Witch, and Lego Black Widow to convey the urgency of what he's just seen. He has explained the threat twice in excruciating detail. As master of the mystic arts and guardian of the time stone, Lego Dr. Strange sees all . . . and he's becoming frustrated that the protectors of Lego Land sitting before him do not share his vision of the obvious. So, he steps back. He considers his audience: individuals representing three generations, three first languages, and extremely varied educational backgrounds. He is explaining data to an audience of varied statistical literacies and shared languages.

DOI: 10.4324/9781003500117-4

So, he takes a deep breath and starts over, this time by summoning a dry erase board. He starts his story again, placing the events he's seen on a timeline, using a different color to represent change over time in his stories, concluding with an image showing three different lines about to converge. On this third telling of the same story, this time he sees a room of wide eyes rather than confusion at his conclusion. Lego Captain America slowly rises from his chair as he engages his com and commands, "Avengers, Assemble!"

Lego Dr. Strange finds that no matter how critical his data findings, his knowledge is next to useless until he can effectively communicate the data. This is why he turns to data visualizations. In this chapter, you will learn:

1. Why we use data visualizations
2. What makes a good visualization
3. How to build data visualizations in Excel and SPSS

Why We Visualize Data

Conveying data to a homogeneous audience is a rare experience. The most likely time for that experience to occur is right now in your classroom, where you have a group of individuals who are likely to be around the same age, speak a common language, and have a similar level of data education. When you leave this classroom, you will need tools to help explain your data to diverse audiences.

Data visualizations are one of the most important tools for clearly conveying data across language barriers and literacy (Kelly et al., 2018; Kelly & Tovey, 2022). The human brain uses half of its resources to process visual information directly or indirectly (Smiciklas, 2012). Because of this, humans jump to understanding visual information before they try to process linguistic information (i.e., spoken, written, or signed words). If the visual data tells a clear story, then most people will start making interpretations of the data and decisions about it without considering the words used to supplement the data.

So, the answer to the question, "Why do we visualize data?" is very simple. We do it because it's the clearest, easiest way to explain data. However, it is important to recognize that while creating data visualizations is easy, given the tools provided for us in SPSS and Excel, creating good data visualizations takes thought and skill.

It is estimated that at least half of the adult population has math anxiety (Boaler, 2012; Sparks, 2020). Math anxiety is stress that some people feel when working with numbers – in any context (Kelly et al., 2022). Math anxiety can happen when taking a math or statistics class, calculating a tip at a restaurant, doing your taxes, or anywhere someone must work with numbers. People who experience math anxiety have a range of symptoms, from physical symptoms, such as a racing heart or sweaty hands, to psychological symptoms, such as trouble concentrating or feelings of self-doubt (Kelly et al., 2020). Math anxiety is not just unpleasant; it is a problem for people when they are trying to work with numbers because their working memory, the same resource they need to manipulate or interpret numbers, is the same resource they need to manage their anxiety symptoms (Morsanyi et al., 2014; Shaikh, 2013). All of this means two things for people with math anxiety: (1) they will want to spend as little time thinking about data as possible because it literally makes them feel bad and (2) the only resource they must work with the numbers is the working memory power left after their symptoms are in check.

You may be wondering why we just digressed to talk about math anxiety in such depth. The reason is that people with **math anxiety** will welcome a data visualization and be glad to take the first message they read from it without having to think too deeply about the data, because it's a way for them to understand the data without triggering their math anxiety. However, that means if the data visualization is not well designed, their cursory glance is going to leave them walking away with the wrong story. Because this applies to half of the population, math anxiety is one of the most important reasons you must think carefully about data visualization designs.

Just as a person's literacy indicates their ability to understand language effectively, their **data literacy** affects their ability to understand data. Data literacy is predicted by how much knowledge someone has about data (e.g., statistical training) and by how much their math anxiety allows them to think deeply about data. Data visualizations are a method of conveying data across data literacy levels.

Test Your Understanding 4.1 – Why are data visualizations useful?

What Makes a Good Visual?

Simply put, a good visual is:

- Accurate
- Clear
- Accessible

It is quite possible for a visual to be only one or two of those things at a time . . . or even none of those things. Consider the pie chart below. A **pie chart** is a method of calculating parts of a whole. An entire anything cannot be more than 100%. Yet, the sum of the percentages represented by Figure 4.1 is 255%. How can something like this happen? What most likely happened is that someone sent out a survey to students on a college campus and asked them whether they believed they would be graduating in four years. Then, when the data were analyzed, 87% of first-year students indicated they would graduate in four years, followed by 62% of sophomores, 85% of juniors, and 48% of seniors. So even though this pie chart is based on accurate data, the presentation tells an inaccurate story.

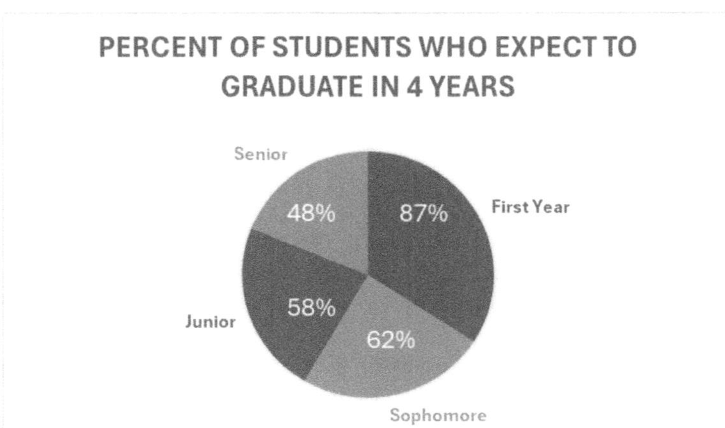

Figure 4.1 Inaccurate pie chart.

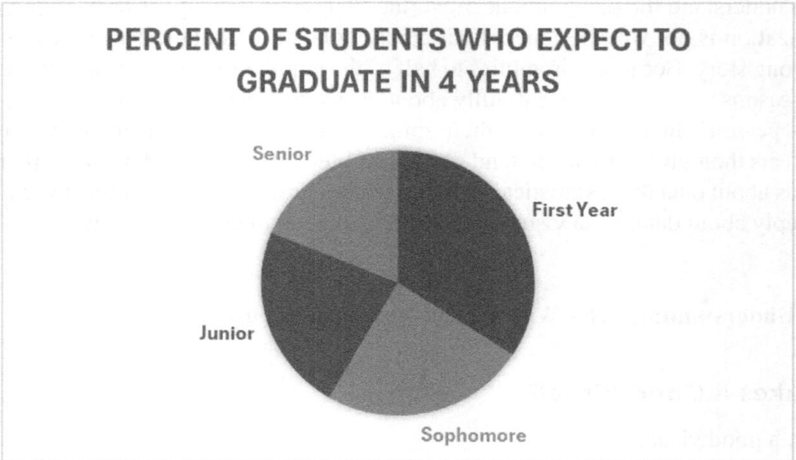

Figure 4.2 Unclear pie chart.

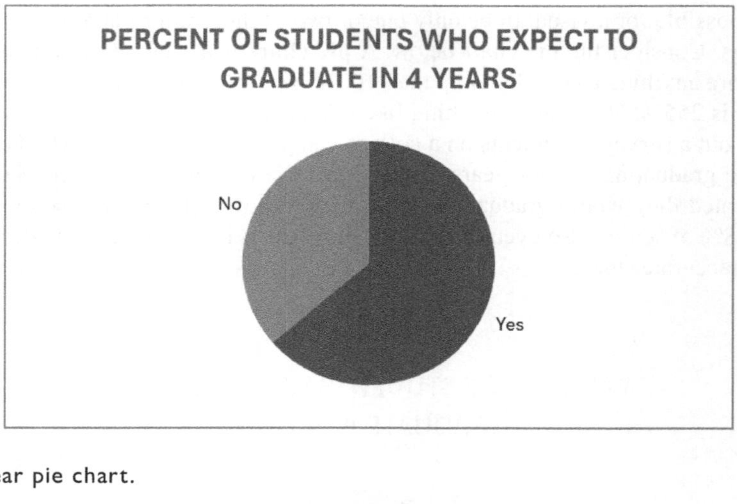

Figure 4.3 Clear pie chart.

Now, take a look at the pie chart in Figure 4.2. The percentage labels that represent what percent of the survey group is represented in each wedge have been removed. So now the pie accurately shows only the percentage of total survey respondents who responded "Yes" to whether they believed they would graduate in four years.

Is the pie chart in Figure 4.2 clear, though? Can you easily see whether the Junior or Senior wedge is larger? Odds are, it's hard to tell just by looking at Figure 4.2. In fact, the human eye is very bad at judging angles (Evergreen, 2020). So bad, in fact, that pie charts are rarely a useful way to convey data if you have more than two wedges. Figure 4.3 is a clear pie chart. With only two wedges, you can clearly see which proportion of the data is larger than the other.

If you need to show more than two parts of a whole, a pie chart is not the right visual for us. In this case, the best option is a **stacked bar chart**. A stacked bar chart is a collection of rectangles representing data quantities that are presented side by side so that the length of the total bar represents the percentage. Most people can compare the size of rectangles far more easily than the size of angles. Figure 4.4 shows what such a visual would look like.

A final consideration is accessibility. **Accessibility** is the extent to which materials are usable to anyone (Drye et al., 2023). When you build visualizations such that some information is conveyed entirely in color, you ensure individuals who have color blindness will not be able to interpret the data as quickly or easily as others. Whenever possible, color should be used to enhance a data visualization, not to be the sole indicator of information. A good practice is to always run your visuals through a colorblindness scanner like Coblis before sharing them with anyone: www.color-blindness.com/coblis-color-blindness-simulator/

When building visuals, there are a few color combinations to avoid that are commonly difficult for individuals with various types of colorblindness to discern:

- Red versus green
- Green versus brown
- Green versus blue
- Blue versus gray
- Blue versus purple
- Green versus gray
- Blue versus gray

Putting any of these two colors side by side in a visual will cause unnecessary ambiguity for individuals with visual impairments. In general, try to alternate warm and cool colors, avoiding the combination of red and green, and use high contrast between colors as often as possible.

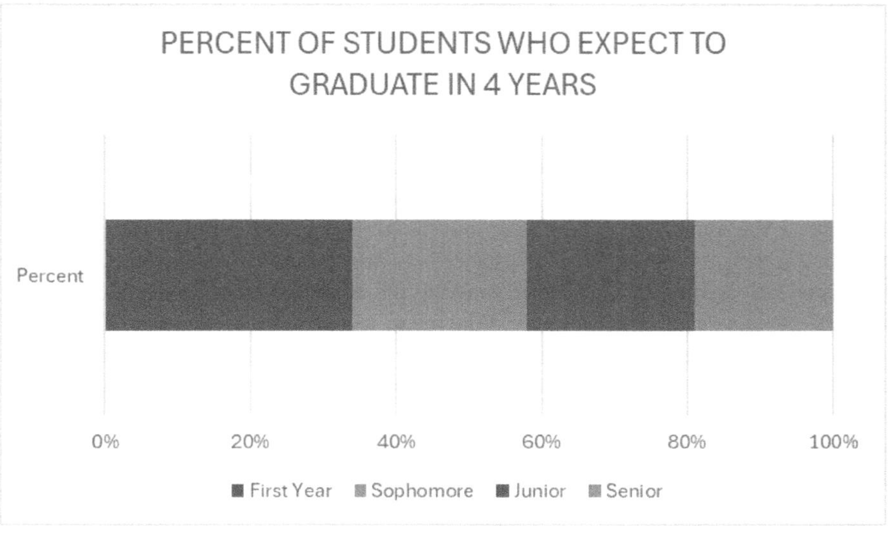

Figure 4.4 Stacked bar chart.

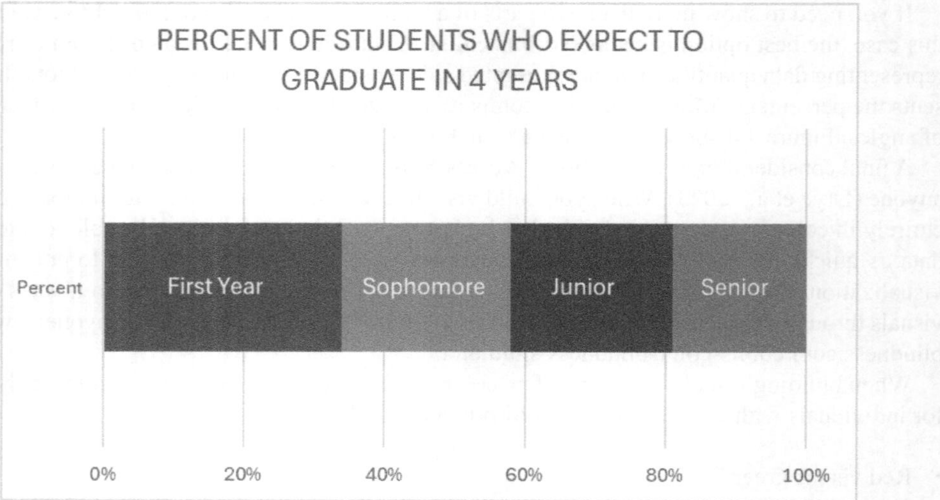

Figure 4.5 An accessible stacked bar chart.

Figure 4.5 shows an accessible version of our data. Here, the bars are labeled, and colors are used only to enhance the visual. At last, in Figure 4.5, you at least have a visualization of the original data that is accurate, clear, and accessible.

Test Your Understanding 4.2 – What does it mean for a visual to be accurate, clear, and accessible?

Visualizing Comparisons

Sometimes you need to compare numbers, showing how they are alike or different, like when an experiment has been conducted. For example, Kelly et al. (2023) conducted an experiment to see how credible students perceived an online instructor based on their Zoom background choice. In this study, it was found that female students showed relatively little difference in their assessment of instructors' credibility based on the instructors' Zoom backgrounds. In contrast, male students assessed instructors as less credible when they used unprofessional backgrounds. This study compared instructor credibility based on students' gender and the instructor's Zoom background to demonstrate how the female data were alike, but the male data differed by background.

In other instances, you want to show the distance between numbers. Croucher et al. (2024) conducted a **meta-analysis** (an analysis of analyses) comparing the mean scores of verbal aggression and argumentativeness across research studies to see if the mean scores on those variables differed by whether the sample used was college students or adult non-students. They were able to show that when it comes to these behavioral tendencies, there were no differences in scores between student versus non-student samples. Thus, this study compared the scores of both groups to show how they are alike.

When it comes to comparing scores (e.g., totals, means, counts), the best options are **bar charts** and **column charts**. These are the same type of chart, using rectangles to show the magnitude of scores on an axis. Column charts have categorical variables on the x-axis and numerical data on the y-axis. Bar charts have categorical variables on the y-axis and numerical data on the x-axis.

Figure 4.6 shows our survey data as a column chart to the left and a bar chart to the right. Note that the bars extend from the categorical data (i.e., class rank). The bars span across numerical data,

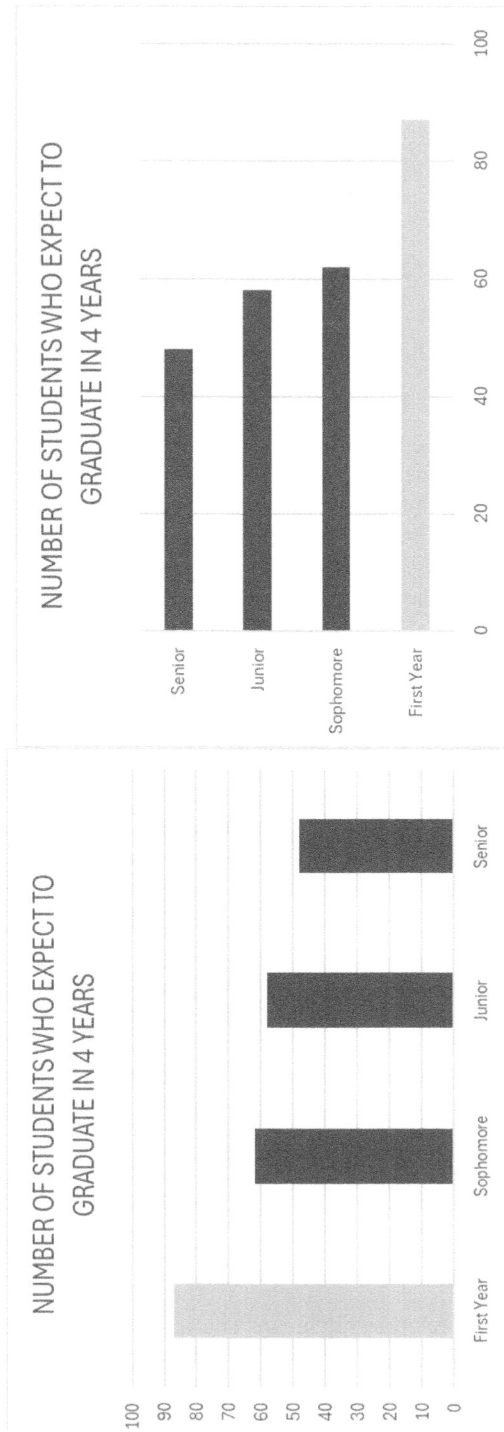

Figure 4.6 Column chart and bar chart.

Table 4.1 Sales Data

Year	*Product A Sales*	*Product B Sales*	*Product C Sales*	*Product D Sales*
2021	$2,500,000	$1,800,000	$1,200,000	$1,000,000
2022	$2,600,000	$1,850,000	$1,300,000	$1,100,000
2023	$2,700,000	$1,900,000	$1,400,000	$1,200,000
2024	$3,000,000	$1,950,000	$1,500,000	$1,300,000
2025	$2,800,000	$2,000,000	$1,600,000	$1,400,000

representing the number of affirmative votes. There are two important design tips to consider when creating bar and column charts:

1. Guiding Lines: Just like the eye has trouble discerning angle size, it also struggles to compare heights across distances without a constant line. Notice that both figures have grid lines every ten points to help you discern the total of each bar. Depending on the story you wish to tell, you may wish instead to insert an average line across the bars, so readers easily ascertain not the total of each bar, but how far the top of each bar is from the average.
2. Color Emphasis: While color should never be the sole indicator of a piece of data, it can be used for emphasis. Notice that your eyes immediately go to the yellow bar because it is different than all the others. This would be a good tool for emphasis if the story you are telling with the data should bring attention to the fact that more first-year students than other classes believe they will graduate in four years.

Visualizing Change Over Time

Like Lego Dr. Strange, sometimes you may need to explain how data changes over time. For this, you may use an area chart or a line chart. A **line chart**, as its name implies, shows data through a line or multiple lines. An **area chart** is a line chart with the space below it shaded. The only time you are likely to choose an area chart over a line chart is when you wish to show parts of a whole.

Consider Table 4.1. This sales data shows a change over time for four products over five years. This data could be presented as an area chart or a line chart, depending on the story being told. Figure 4.7 shows the line chart to the left and the area chart to the right. The data shows the exact same information over time, but in a different format. The line chart emphasizes how individual product sales trend over time. The area chart shows how the total sales change each year and emphasizes the role each product played in the total sales.

Test Your Understanding 4.3 – When would you choose to use an area chart over a line chart?

Visualizing Relationships

To say one wants to visualize a relationship means one wishes to show how consistently the magnitude of one continuous variable aligns with another. This is really the idea behind correlation and regression, which you will not read about until Chapter 8. You can think about them on a very basic level for visualization, though.

Table 4.2 shows data from which students assessed how clear they felt their professor was throughout the semester and how much burnout they feel at the end of the semester. (For anyone reading

Table 4.2 Burnout and Clarity

Burnout	*Clarity*
7	4
7	7
7	1
6	5
2	2
1	4
7	4
7	1
7	4
6	1
6	2
7	1
5	5
5	6
7	1
7	4
7	5
7	6
4	7
7	2

this with a measurement background, please rest assured that this data has been rounded to the nearest whole number for this introduction visual, but the original data was collected with appropriate measures.) If you wish to know whether these data have a relationship, you wish to know how well you can predict the value of burnout by knowing a student's assessment of clarity or vice versa. To ascertain this, look at a **scatterplot**, a visual that plots continuous data as coordinates on a Cartesian plane using one variable's values for x and the corresponding variable's value per participant for y.

Figure 4.8 shows the scatterplot of this data. Notice the first row of data in Table 4.2 has a 4 for burnout and a 7 for clarity. Now, notice in our dataset that the coordinate (4, 7) is marked on our scatterplot (and circled to make it easier to find). Each row of data is plotted in this chart. Because these data have been rounded, they are not appropriate for making inferences. You'll get to that in Chapter 8 with authentic data. However, when you look at a scatterplot, you are looking to see if there is a pattern in how the data is plotted (i.e., the variables are related) or whether the data is positioned at random.

Data Visualizations in Excel

No matter the chart type you want to build in Excel, the process is the same. As suggested in Chapter 3, you will have the greatest ease in Excel if you format your data as a table. Once you've done this:

1. Highlight the data
2. Click on the Insert ribbon
3. Choose the chart type you wish to have from the available icons

The location of these buttons is shown in Figure 4.9.

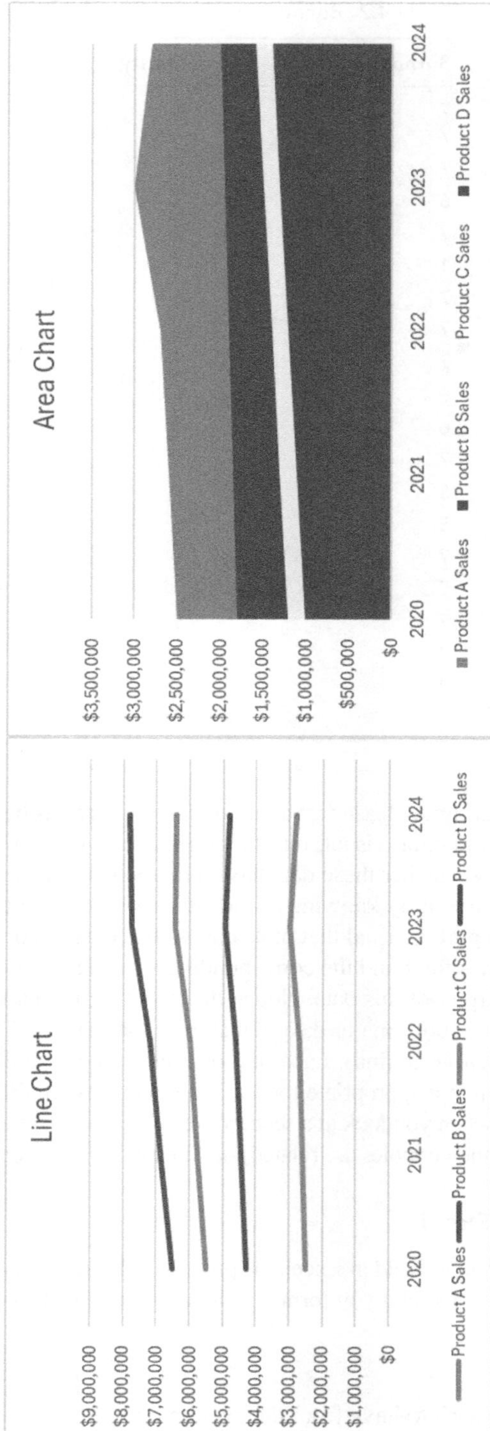

Figure 4.7 **Line versus area chart.**

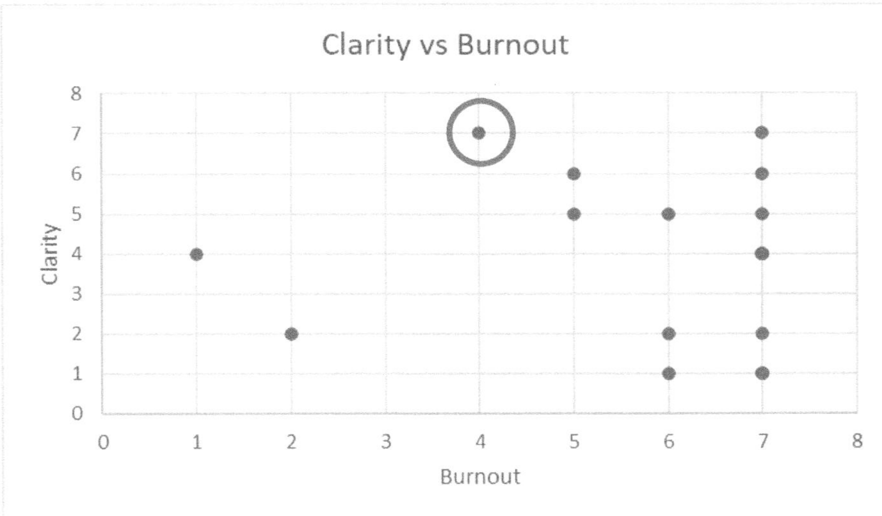

Figure 4.8 Scatterplot.

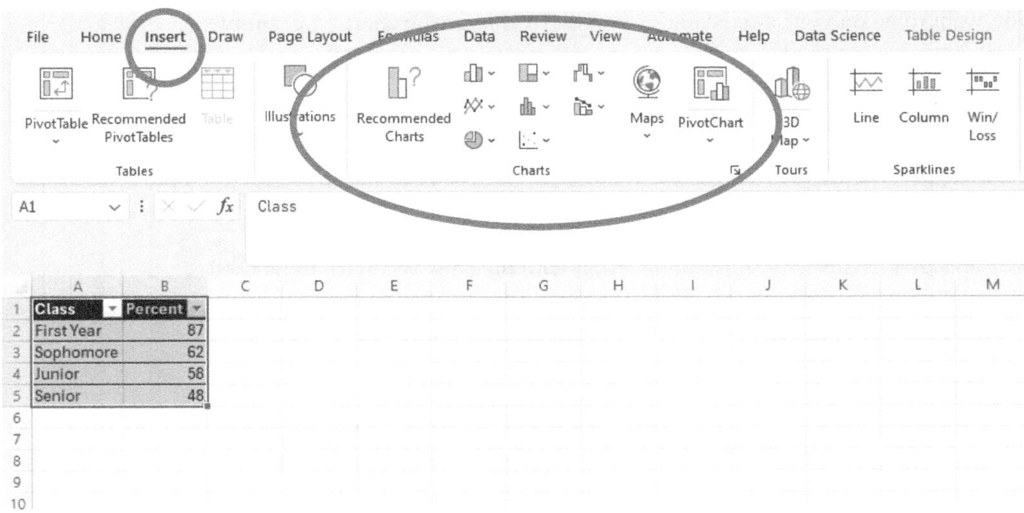

Figure 4.9 Insert tab.

Often, you can select the Recommended Charts option and be presented with the visualization most appropriate for the data. Sometimes Excel does not know what is best for the user, though. In these cases, you may need to modify your chart. Excel allows you to change the chart type or aesthetic qualities of the chart at any time.

Once you have inserted a chart in Excel, the first step to editing it is to click on it. Once you've done this, you will have two new ribbons in Excel called Chart Design and Format (see

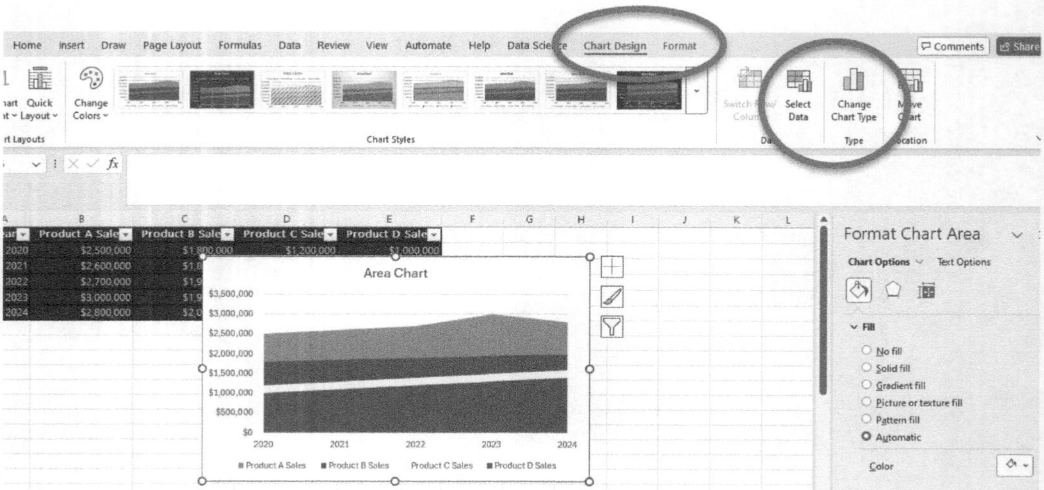

Figure 4.10 Chart Design and Format tabs.

Figure 4.10). The Chart Design ribbon gives you options to change the chart type or to tell Excel that it's reading the data wrong. It is very common to have to click on Select Data to tell Excel what you want to be on your x-axis and/or y-axis. To change the chart type, simply:

1. Click on Change Chart Type
2. Select the type of chart you prefer

To change which information Excel is using to build axes and plot:

1. Click on Select Data
2. Replace the cells chosen for your visual with the cells you wish to be used

It is also easy to change the aesthetic features of your charts. To do this:

1. Click on the Format ribbon
2. Use the dropdown menus to change font, add text boxes, change colors of the background, and many other small aesthetic tweaks, similar to the way you format content in Word or PowerPoint

Another way to modify charts in Excel is simply by double-clicking on what you want to be changed. If you want to change the title, double click it. If you want to change the color of one element in your chart, double click it. Excel assumes that if you double click on something that you are attempting to edit it and will take you directly to the menu options for altering that element.

Clicking on the chart also pulls up shortcut menus to open windows to edit small pieces of the chart. Figure 4.11 shows the menu shortcuts that appear once a chart is clicked on. The plus sign provides you with the chart elements you can change (e.g., axes, labels). The paint brush provides you with preset color palettes you may choose from to alter your chart. The funnel allows you to choose to add or remove variables from your visual, just as the filter allowed you to change your tables in Chapter 3.

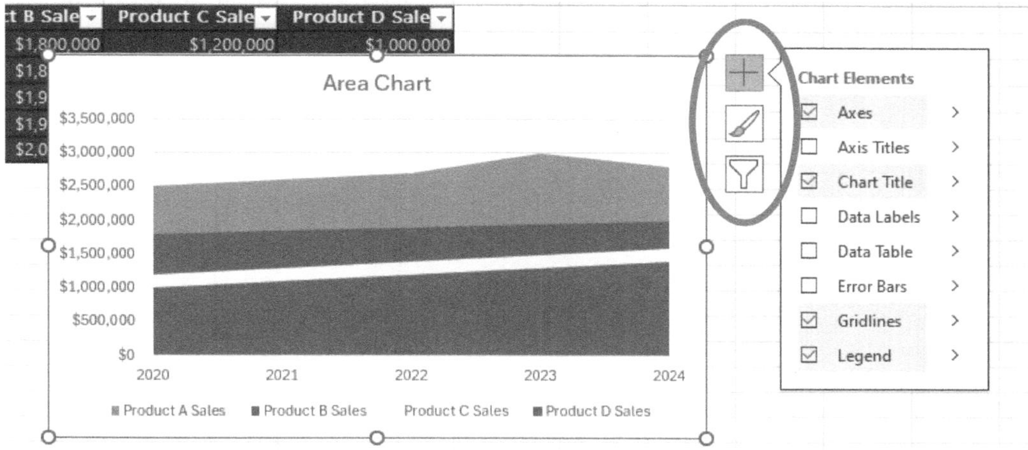

Figure 4.11 Changing an Excel chart.

Data Visualizations in SPSS

If you need to visualize data, the most important thing for you to understand about doing that in SPSS is that SPSS was never designed to do this job. SPSS has been retrofitted to allow for some data visualization, but the customization is limited, and the details are poor. Always use Excel for building data visualizations over SPSS.

That said, if you must build a data visualization in SPSS for some reason, here are the steps to do so. As shown in Figure 4.12, use these steps to begin building a data visualization in SPSS:

1. Click on Graphs
2. Click on Chart Builder

Once you have the Chart builder tool opened:

1. Click on the type of chart you wish to build in the Gallery (see Figure 4.13)
2. Drag the icon that represents the way you wish your chart to look into the chart preview window
3. Once the chart type is dragged into the chart preview window, you will be given space to drag your variables to elements of the chart, such as the x-axis and y-axis. You can do this from where they are listed to the left of the chart preview window
4. After you have dragged variables to their space in SPSS, you may make edits to color in the Chart Appearance
5. Click OK

There are a few things that can be changed about a visual in SPSS. However, it is quite capable of making simple visualizations, especially with three or fewer variables.

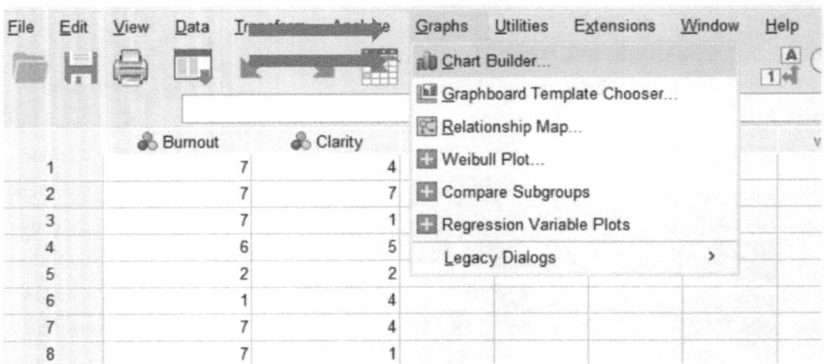

Figure 4.12 Beginning a build in SPSS.

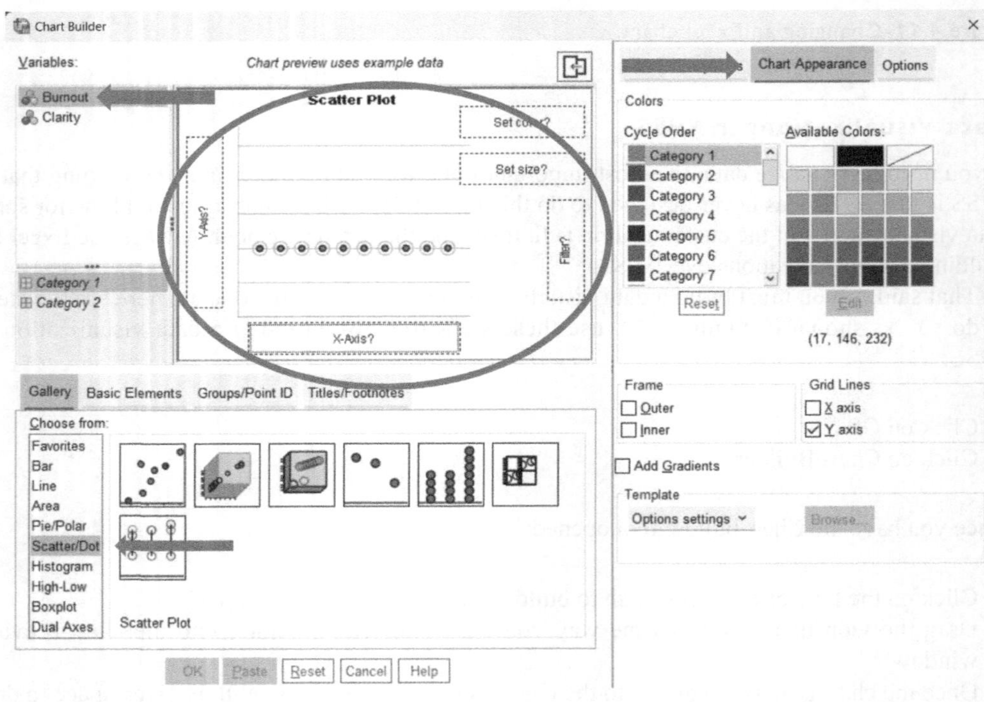

Figure 4.13 Chart Builder.

Summary

Just like Lego Dr. Strange sought the aid of data visualization to convey his point, you, too, will need to rely on the power of data visualization to overcome anxiety, language, and literacy barriers. The essence of effective visualization lies in its accuracy, clarity, and accessibility. While both Excel and SPSS offer visualization tools, Excel is going to be your most adaptable and strongest tool for this job. In concluding this chapter, it is important to mention that the strength of data visualization lies not in complexity but in its ability to communicate insights with precision and universality.

References

Boaler, J. (2012). Timed tests and the development of math anxiety. *Education Week*. www.edweek.org/teaching-learning/opinion-timed-tests-and-the-development-of-math-anxiety/2012/07?tkn=VOQFK9Wqv%2FQyFHqgUJA20N6MixLTNTr2KA%2Fy&cmp=clp-edweek

Croucher, S., Kelly, S., Jackson, K., Nguyen, T., & Elers, P. (2024). Does student sampling impact our understanding of argumentativeness and verbal aggressiveness. *Argumentation and Advocacy, Early Access*. https://doi.org/10.1080/10511431.2024.2390775

Drye, S. L., Kelly, S., & Woodard, T. (2023). Professionals' understanding of accessibility regarding business communication materials. *Business and Professional Communication Quarterly*, *86*(3), 235–256. https://doi.org/10.1177/23294906221133068

Evergreen, S. (2020). *Effective data visualization: The right chart for the right data*. Sage.

Kelly, S., Croucher, S., Kim, K. Y., Eskicorapaci, N., Permyakova, T., Turdubaeva, E., Rocker, K. T., Stanalieva, G., Orunbekov, B., & Rimkeeratikul, S. (2022). A general math anxiety measure. *Education Sciences*, *12*(6). https://doi.org/10.3390/educsci12060370

Kelly, S., Croucher, S., Sandel, T. L., & Yu, F. (2018). Infusing infographics into the business curriculum: A study of infographic interpretations in China, New Zealand, and the United States. *Global Advances in Business Communication*, *7*, 1–21. https://commons.emich.edu/gabc/vol7/iss1/3/

Kelly, S., Kim, J., Goke, R., & Bryant, K. (2023). Does instructors' choice of virtual background in online classes matter? Effects of personal vs. professional virtual background use. *International Journal of Human-Computer Interaction*, *40*(22), 7314–7321. https://doi.org/10.1080/10447318.2023.2263695

Kelly, S., Romero, A., Morrow, J. A., Denton, Z., & Ducking, J. (2020). Instructor misbehaviors and math anxiety. *Communication Reports*, *33*(1), 27–40. https://doi.org/10.1080/08934215.2019.1675737

Kelly, S., & Tovey, T. L. S. (2022). Data visualization as communication: The role of business communication in data analytics. *Business Communication Research and Practice*, *5*(1), 42–45. https://doi.org/10.22682/bcrp.2022.5.1.42

Morsanyi, K., Busdraghi, C., & Primi, C. (2014). Mathematical anxiety is linked to reduced cognitive reflection: A potential road from discomfort in the mathematics classroom to susceptibility to biases. *Behavioral and Brain Functions*, *10*, 31. https://link.springer.com/article/10.1186/1744-9081-10-31

Shaikh, S. N. (2013). Mathematics anxiety factors and their influence on performance in mathematics in selected international schools in Bangkok. *Journal of Education and Vocational Research*, *4*, 77–85.

Smiciklas, M. (2012). *The power of infographics: Using pictures to communicate and connect with your audience*. Que Publishing.

Sparks, S. D. (2020). The myth of fueling math anxiety. *Education Week*. www.edweek.org/teaching-learning/the-myth-fueling-math-anxiety/2020/01

Chapter 5

Describing Data

What Will I Learn in This Chapter?

In the bustling realm of Lego Land, chaos reigned supreme, with bricks of all shapes, sizes, and colors littering the landscape. Taking a stand in all the disorder stood Lego Captain America, a symbol of order in a landscape of creative possibilities. Before Cap could begin bringing order to the world, he needed to understand what the data, or in this case, the Lego bricks, looked like. Just as one needs to glance at data before analyzing it, Lego Captain America needed to survey what lay before him before he could begin his build.

Lego Captain America began to organize the bricks in ways that would allow him to sort, categorize, and understand what lay before him. Using the mean, he arranged the bricks based on their average size, identifying those that would serve as sturdy foundations for towering structures or delicate details in mosaics. The median guided him in placing bricks of similar sizes together,

DOI: 10.4324/9781003500117-5

ensuring balance and stability. With the mode, he identified the most common colors, giving him the basic information he needed to dream up vibrant patterns. Armed with an understanding of what was available, Lego Captain America transformed the chaotic landscape into a metropolis of order and beauty. Understanding how to describe data, just as Lego Captain America understood the Legos available to him, enables individuals to make sense of the information they encounter. Whether it's organizing Lego bricks or analyzing exam scores, having the ability to describe data is the first step in understanding what you have and what you can do with it. In this chapter, you will learn:

1. How to think about the center of your data
2. Other ways to visualize your data
3. How to calculate indicators of your data's shape in Excel and SPSS

Literally Looking at the Data

There are several different ways you can take a glance at the data. As we practice looking at this data throughout this chapter, we'll be using the example of a set of 50 exam scores. (This data was created for this example.) Here is what the created exam data looks like:

Table 5.1 Exam Data

100%	89%	82%	76%	71%
97%	88%	82%	76%	71%
96%	88%	82%	75%	71%
95%	88%	82%	74%	71%
94%	88%	80%	73%	70%
92%	87%	80%	73%	69%
92%	87%	79%	73%	69%
92%	86%	79%	72%	68%
92%	85%	78%	72%	67%
92%	84%	77%	72%	67%

Sometimes it's helpful to understand data by making a **frequency distribution**. A frequency distribution is a list of all values that appear in your dataset and a count of how many times that value appears in the dataset. For example, 100% appears once, but 92% appears five times. As visual creatures, though, humans are going to understand and see frequency distributions of data more quickly if that information is presented visually. The easiest way to do this is through a histogram.

A histogram plots the number of times a value appears in your dataset. Histograms use interval notation on their x-axis to represent values included in a bar. Thinking back to your math class, you may remember using parentheses and brackets to explain where values of x exist in a graph. The parentheses mean ranging to but not including. The brackets mean ranging to and including. The second leftmost column has the values (68%, 69%] shown below it. This means that the bar represents the number of exams that include values infinitely close to but not exactly 68% and going all the way up to 69%. Exam values of 68% are counted in the bar to the left.

The interval length of a histogram is called the **bin.** You can set the size of your bin to whatever makes sense to you. Often, the smallest distance between data points is a reasonable bin width. Here, the bins are set in ranges of 1% so you can see one bar per score. Another perfectly valid way

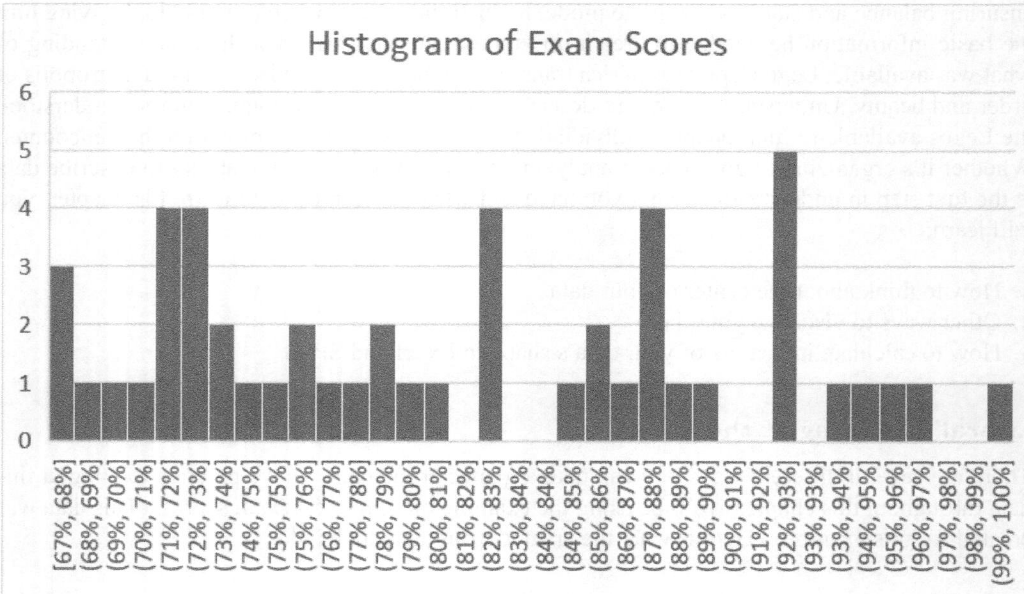

Figure 5.1 Exam histogram.

to show this data would be to split the bar into bins the size of the range between letter grades. That would give one bar for A's, one for B's, one for C's, and one for D's (assuming the same range of values between letter grades).

The purpose of a histogram is to allow you to have a quick overview of common data values and holes in the data. Here, we see several holes in the A and B range (again, assuming each letter grade spans ten points), but no holes in the D and C range. Much of what we are looking at in a histogram answers questions about relationships from the data's center, our measures of central tendency.

Measures of Central Tendency

The term **measures of central tendency** refers to the collection of tools we can use to figure out where the center of the data is and where another data point is in relation to that center.

The first measure of central tendency is the **mean**. The mean is also called the "average." We solve for the mean using the following formula:

Formula 5.1: Mean of Population

$$\mu = \frac{\Sigma x}{N}$$

The lowercase Greek letter mu (μ) is used to represent the average score in a population of data. In the formula, the letter x means a value in the dataset. The capital Greek letter Sigma (Σ) means add up every item that fits into the formula that follows. The capital letter N means the number of items in the population. So, when what follows Σ is simply x, that means add all the x's together. $\frac{\Sigma x}{N}$ then means to add all your values in your dataset and divide it by how many values you have in the population set. For the dataset here, that would mean adding all 50 values and dividing by 50:

$$\frac{100\% + 97\% + 96\% + \ldots + 67\%}{50} = 81\%$$

Notably, the formula for the mean and sample are identical in practice. The only difference is that the sample formula uses a lowercase n to represent that the data is from a sample rather than a population, and $\bar{x}$ represents the mean of a sample.

Formula 5.2: Mean of Sample

$$\bar{x} = \frac{\Sigma x}{n}$$

A second measure of central tendency is the **mode**. Mode is the value (or thing) that appears most frequently in a dataset. In the exam dataset, the mode is 92%. It is the only number that appears up to five times. Modes can also be considered qualitatively. For example, imagine you have the grocery list below to guide your shopping. The mode in this data is yogurt because it is the item you have selected to buy in the greatest quantity.

Groceries

- 1 gallon milk
- 1 carton eggs
- 7 yogurt (8 oz)
- 2 bags granola
- 1 box lasagna noodles
- 2 blocks Monterey Jack cheese
- 2 jars spaghetti sauce
- 1 container ricotta cheese
- 1 bag grated Parmesan cheese

A third measure of central tendency is the **median**. The median is the middle number in a dataset when the data is organized in ascending or descending order. (Our exam data is organized in descending order, but we could switch it to smallest to largest to put it in ascending order.) If the dataset has an odd number of datapoints, then there is just one number in the center, and that is your median. If the dataset has an even number of datapoints, then the average of the two most central datapoints is the median. Here, our data has 50 datapoints, so our median is the average of the two values listed in positions 25 and 26. Both of those values are 80%. So, our median is:

$$\frac{80\% + 80\%}{2} = 80\%$$

So, why do we need mean, median, and mode? Which is the best? Which should you use? The answer is every student's least favorite: it depends.

Sometimes what you need most is the mode. Imagine you are a student worker for a department on campus, and that department just got a small amount of money refunded to their budget (perhaps $50) that they need to spend or lose in the next few hours to close budgets for the year. You have now been tasked to look at the data on how office supply monies were spent this semester to figure out how to best spend that $50. Are you going to uniformly order the same number of items (e.g., pens, pencils, paper, markers) until you run out of money? Or are you going to look at the data indicating which item gets used the most frequently and buy a few extra of that most coveted item? A wise person would buy backups of the most frequently used item, which means using the mode to understand your data.

Sometimes what you need is the median. Imagine you've received an exam score and are devastated to see that your score is a low D – 61%. Then, you are even more devastated that the class average is a B+. But then you learn the median score on the exam was 68%. Now, how do you feel? What should you do with that information? If the average is a B+, but the median score is close to your own score, that means you fell close to the middle of the pack. We expect the median to be somewhere in the C range for a well-designed test. This means there are a few students in the class who earned outstanding scores, and they've pulled the average up to where it doesn't represent most student scores. This means there is a reason to have a conversation about the nature of the exam and how to prepare for it. What did those couple of students with outstanding scores do differently? Perhaps they used a study reference you didn't know existed but can on the next test, so you're well prepared for the next test. Perhaps those students cheated, which ruined the curve for everyone, and you need to make an academic integrity report. Perhaps those few students have had a prior class that taught them how to think about the exam material in a way that students without that preparation would not, in which case, the wording of the exam needs to be discussed with the faculty, and perhaps a change in pre-requisites for the class.

Sometimes what you need is the average, though. For example, knowing the average number of absences that students who earn an A in a class have tells you how often you should be in class. If the average student who earned an A only missed one day of class, you shouldn't plan to miss more than one, unless you're okay with a lower grade. If the average student who earned an A spent three hours outside of class studying and doing homework for this class each week, that tells you should plan to spend at least three hours working on the class outside of class each week if you want an A. All these measures are informative. Sometimes one is better than the other at telling you what you need to know. Often, you'll find it is most useful to glance at them all.

Test Your Understanding 5.1 – How much money do you think students in your class earned on average during their summer break? What do you think the mode for the class's summer income was? What do you think the median for the class's summer income was?

When you read social science publications, the first place you see central tendencies is typically in the Method section, where researchers describe their sample. Zeng et al. (2020) published a study on employees' intention to engage in whistleblowing at work. In their method section, they described their sample using the mode and mean. They shared that the average age of their participants was 38.8 years old. They shared that the mode of participant gender was female versus male. They also shared that the mode of organizational tenure was the group of participants who had worked for the organization 3–5 years versus those who had been there longer or for less time. All social science papers, whether they be qualitative or quantitative, typically use these measures of central tendency to describe their sample.

Measures of Variability

While measures of central tendency are concerned with where the middle of the dataset is, measures of variability deal with how spread out the data are. Measures of variability are needed in addition to measures of central tendency to get a better understanding of what the data look like. For example, imagine three boxes whose average weight is 140 pounds. That could be three boxes that weigh 140 pounds each; three boxes that weigh 145, 140, and 135 pounds, respectively; or two boxes that weigh 100 pounds and one that weighs 220 pounds. The three boxes that weigh 140 pounds have no variability in their weight because each box weighs the same. The boxes that all weigh within 5 pounds of 140 have little variability in their weight. The boxes that weigh 40 to 80 pounds different from the mean have considerably more variability than the other sets. Centrality alone does not paint a clear picture of data.

The first measure of variability is **range.** Range is the difference between the largest and smallest values in the data. In our exam score example from earlier, this would be 100% – 67% = 33%. That means the range in the data is 33%. While this tells you the difference between the best and worst score from the exam, the range does not always tell a complete picture. For example, if someone has scored a 10% on the exam, a range of 90% would imply a lot more variability in scores than is truly present because of one failing grade.

This is why sometimes it's handy to look at a subset of ranges in a dataset. **Interquartile range** is the difference between the lowest and highest 25% of the data. In the example of our exam data, the lowest 25% of data is the lowest 12.5 exam scores. As there is no such thing as half an exam score, we round and accept 13 scores to represent the lowest quartile. That makes the highest score in the lower quartile 72%. The middle second quartile cut-off is at 50%, or the lowest 25 exam scores, making the second quartile limit 80%. The third quartile falls at the top of the lowest 75% of exam scores, or the lowest 37.5 exam scores, which we round to 38. This means the third quartile cut-off is at 88%.

In statistical notation, it is common to notate quartiles with an italicized capital *Q*. We would then say that the values associated with the quartile ranges for our dataset are:

$Q1 = 72\%$ $Q2 = 80\%$ $Q3 = 88\%$ $Q4 = 100\%$

As such, the interquartile range is $Q3 - Q1 = 88\% - 72\% = 16\%$. This gives us an idea about how spread out the middle 50% of exam takers were. The 50% of exam takers who scored in the middle range of scores on the exam were spread out by 16% overall. This tells us how close people were to the middle when they were in the middle 50% of exam scores. This describes part of the data, but it suffers from the opposite limitation of range: it doesn't give us any idea about outliers.

Table 5.2 Exam Data

100%	89%	82%	76%	71%
97%	88%	82%	76%	71%
96%	**88%**	82%	75%	71%
95%	88%	82%	74%	71%
94%	88%	80%	73%	70%
92%	87%	**80%**	73%	69%
92%	87%	79%	73%	69%
92%	86%	79%	72%	68%
92%	85%	78%	**72%**	67%
92%	84%	77%	72%	67%

Formula 5.3: Interquartile Range

$Q3 - Q1$

As a social scientist, the measures of variability you will find most useful are standard deviation and variance. Before we talk about standard deviation, let's take a moment to think about what those words mean separately. A deviation is a departure. Standard means normal. That means a standard deviation is how far it is normal to depart from something. In terms of statistics, the **standard deviation** is how far removed a single datapoint is from the center of a dataset. The formula for the standard deviation of a sample is:

Formula 5.4: Standard Deviation of Sample

$$S_x = \sqrt{\frac{\Sigma(x-\bar{x})^2}{n-1}}$$

Again, x is any value in the dataset. $\bar{X}$ is the mean. Then n is the total number of items in the sample. The formula for the standard deviation in a population is similar:

Formula 5.5: Standard Deviation of Population

$$\sigma_x = \sqrt{\frac{\Sigma(x-\mu)^2}{N}}$$

There are only two differences in this formula. First, note we use s to represent the standard deviation of a sample, and the lowercase Greek letter sigma (σ) represents the standard deviation of the population. Also, note that in the formula for sample, the denominator is the number of items in the sample minus 1. For the population formula, the denominator is the total number of items in the population. This difference is to accommodate for a bit of measurement error in the sample, as it is unlikely to be a perfect representation of the population.

For our example with the exam data, we have the total number of exams in the population. Because of that, we would find the normal distance from the center of scores, the standard deviation for the exams, using the population formula.

$$\sigma_x = \sqrt{\frac{\Sigma(x-\bar{x})^2}{N}} = \sqrt{\frac{(100-80)^2+(97-80)^2+...+(67-80)^2}{50}} = 10.3$$

How is one to interpret a standard deviation of 10.3 in this case? It means most scores fall within $\pm$ 10.3% of the mean. Why is this useful? Well, there are several things you can glean from this. First, someone who scores more than 10.3% above the mean did exceptionally well or exceptionally badly because the standard deviation gives you a sense of normalcy. Standard deviation may tell you more, but we cannot talk about it yet because we first must talk about variance and distributions. Hang in there, though! If you keep reading, you'll find out other reasons this information is useful.

The last measure of variance we'll talk about is the statistical term "variance." **Variance** is literally the standard deviation squared. Thus, the formulas are as follows.

Formula 5.6: Variance of Sample

$$S^2_x = \frac{\Sigma(x-\bar{x})^2}{n-1}$$

Formula 5.7: Variance of Population

$$\sigma^2_x = \frac{\Sigma(x-\bar{x})^2}{N}$$

We're not going to show you how to fill in the formula; if you understand the standard deviation formula, you understand you do the same thing for variance, except you skip the last step of taking the square root.

Why is variance important? Variance is the best indicator of how spread out the data is. The variance is not as easily influenced by outliers as the range and it's not subject to over-relying on the middle of the pack like the interquartile range. It's also easier to work with by hand since it does not involve the square root, which creates irrational numbers in formulas. The larger the variance, the more spread out the data, but the smaller the variance, the more tightly compacted the datapoints are.

Test Your Understanding 5.2 – Think about what you estimated about your class's summer income. How spread would that data be?

Distributions

Now we're going to talk about something amazing. In nature, if you take a large, random sample of anything, and then make a histogram of it, that histogram is going to take the shape you see in Figure 5.2.

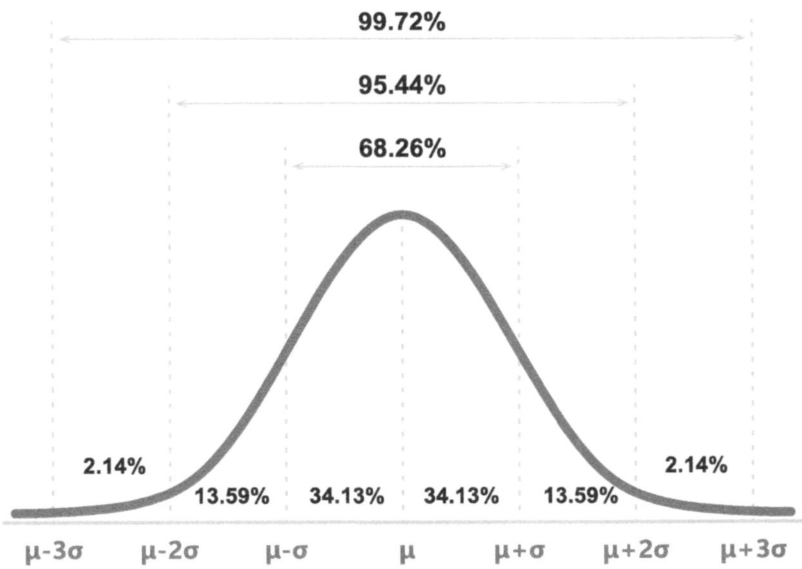

Figure 5.2 Normal distribution curve.

This curve is called the **normal distribution.** You may have also called this shape a "bell curve." In nature, we always see the greatest number of datapoints closest to the mean. This can be any data! You can talk about real exam scores in a class, the height of all people in this country, or the public speaking anxiety scores of people at your school. Odds are, if you get that data and plot it, it's going to look something like this curve.

If you look back at our histogram of exam scores, that histogram shape doesn't look anything like this: it has lots of peaks and gaps. As a data examiner, when you see a histogram that looks so different from this curve, you should get suspicious. Something is wrong with that data. In our case, we already know what was wrong with it – the data is fake. It was built by a computer program. First, the computer was told to randomly derive a list of numbers between 1 and 100, but then it was told to multiply each number by 4, 3, or 2 until it was impossible to multiply it again without going over 100. Those are not real scores, and the distribution should tell you that it looks fake.

In statistics, we get a lot of information from the normal distribution. In Figure 5.2, you'll notice there are four shaded areas. The darkest portion of the distribution shows the area that is one standard deviation away from the mean. 68% of all data in a population, and in a large random sample, will fall within this part of the histogram. 95% of the data will fall within two standard deviations of the mean. 99.7% of the data will fall within three standard deviations of the mean.

When you look at the distribution of the data, if the smoothed shape of the histogram looks slightly off, then it's usually an indication that you have a sampling error, that your sample did not match the population. When the distribution leans to one side, this is called **skewness**. If the curve leans to the right, meaning that most people have scores lower than the mean, this is called being **skewed left**. If the curve leans to the left, meaning most people have scores greater than the mean, this is called being **skewed right**. When a distribution is skewed, that means there are some extreme values making the shape of the distribution lean to one side or the other (i.e., skewing the mean). These outliers should be examined to see whether they should be dropped from the sample before further data analysis. Look at the types of skew you may encounter in Figure 5.3. Also, notice that in a perfect, normal distribution, the mean, median, and mode are the same!

If the curve is too tall or too short, this is called **kurtosis.** Very tall, skinny curves are called **leptokurtic**. Short, flat curves are called **platykurtic**. Take a glance at Figure 5.4 to make that a bit clearer. When there is kurtosis in data, this typically means that something has interfered with the data collection. For example, if the mean score on an exam is 98%, that means there are a lot of students scoring somewhere between 96% and 100%. That is going to give a very tall, very

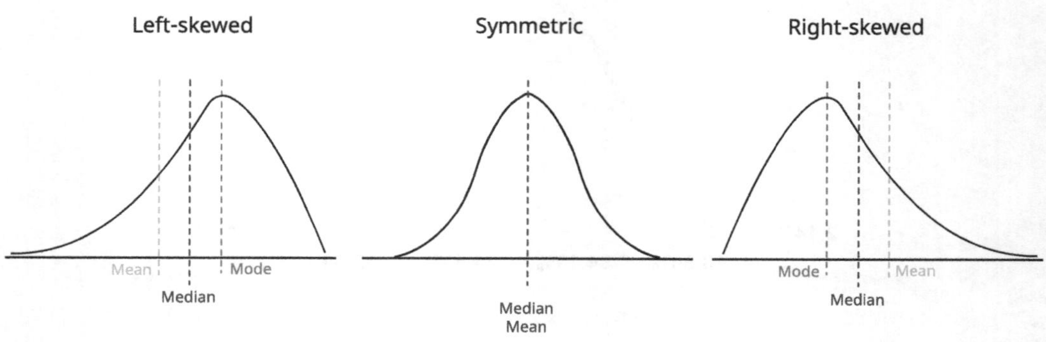

Figure 5.3 Skewness.

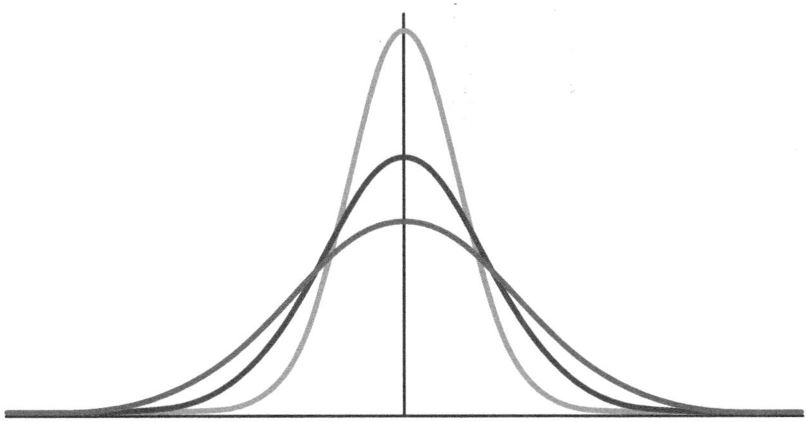

Figure 5.4 Leptokurtic and platykurtic.

thin distribution. It means either that the test was VERY easy (a problem with the data collection instrument) or that the test was given to far above average students (a problem with the sampling technique). Either way, examining these curve shapes tells us a lot about the data.

Test Your Understanding 5.3 – Think about what you estimated about your class's summer income. Imagine that your favorite movie star showed up to class today. How would that change the central tendencies of the summer salaries of the people in class? How does having someone with such a large income added to the data skew your distribution?

Both skewness and kurtosis have a formula for calculation. While the formula for skewness is manageable, kurtosis is a bit overwhelming. So, we're not including those here, but feel free to Google the formulas if you're extremely curious. At the end of the chapter, you'll see how to get Excel and SPSS to make those calculations for you. What we want you to remember about skewness and kurtosis is that when you look at the skewness and kurtosis values for your data, you should be concerned if the value is less than −2 or greater than 2. That means something strange is happening with your data. You'll learn more about this strangeness and what to do about it in Chapter 6.

For social scientists, here is an important factoid to remember about normal distributions: You need at least 20 datapoints to have enough data to form a normal distribution. If you see data analyzed in which a subset of the data being considered has less than 20 values, that's a big red flag. Most of the analyses you'll learn in this book assume that your data is normalized. If that assumption is violated, your output is nonsense. We'll remind you of this factoid again later because it is a critical factoid to memorize as you analyze data, particularly experimental data.

Earlier, we discussed that all social science papers tend to use measures of central tendency to describe their study sample. Quantitative papers typically always use additional descriptive statistics later in the paper to describe the measures they use. For example, Croucher et al. (2021) compared argumentativeness behaviors across individuals who live in France, Germany, Nicaragua, Philippines, Turkey, the United Kingdom, and the United States. Before testing any of their hypotheses to compare argumentativeness across these countries, they reported the mean, standard deviation, range, skewness, and kurtosis for that measure within each country. They reported the mean, standard deviation, and range to help readers imagine what the data looks like. (Did the scores tend to be uniform in a particular culture or varied? Did they tend to be high or low in certain countries?) They reported the skewness and kurtosis so that the readers knew that the data had a normal distribution in each country.

Z-scores

A final way to imagine the shape of data is by transforming datapoints into *z*-scores. A ***z*-score** is a transformed value for any datapoint with respect to the mean and standard deviation of the population, which can be used to determine placement on a distribution. The formula for *z* is as follows:

Formula 5.8: Population *Z*-Scores

$$z = \frac{X - \mu}{\sigma}$$

Here, X is a value in the population, μ is the mean of the population, and σ is the standard deviation of the population. In the back of your book, you'll find the Z Table in Appendix Z. The Z Table shows you which percent of the data is smaller than the data with your *z*-score.

For example, imagine that you have a dataset for a population. This dataset shows the public speaking anxiety score on a scale of 1–7 for every single undergraduate at your school. The mean public speaking anxiety score is 5.8 with a standard deviation of .6. If we want to find the *z*-score for someone with a public speaking anxiety score of 6, then we solve as follows:

$$z = \frac{X - \mu}{\sigma} = \frac{6 - 5.8}{.6} = \frac{.2}{.6} = .33$$

Now we look at the Z Table to see what proportion of data a *z*-score of .33 is associated with. Table 5.3 shows the snippet of information you need. You look for your *z*-score under the column labeled "*z*." To the right of that, you have the "Left" column. The value in the left column tells you what percentage of the data falls under your z-value. The "Right" column tells you what percentage of your data falls above that *z*-value. You interpret this as 63% of the students had a score lower than 6, and 37% had a score higher than 6. In the Z Table in the appendix, the numbers listed tell you what percentage of the data falls to the left of (below) that *z*-score.

Now, you might be wondering when you would ever need to do this when you'll be working with datasets small enough that you can quickly have Excel, or some other program, organize the data in ascending order, look to see how many values you have greater than 6 (or whatever number you're curious about), and quickly calculate the percentage without first calculating the mean and standard deviation. Here's the truth – the audience this book is written for will likely rarely need to calculate *z*-scores. This is something much more useful in the natural sciences and psychology.

However, you do need to understand what a *z*-score is, and here's why – by the time you get to the back of this book, a lot of things you'll be looking at to make inferences will have been

Table 5.3 Z Table Snippet

Z	*Left*	*Right*
.33	.6293	.3707

transformed into z-scores. The lovely thing about z-scores is that they are consistent with the normal curve. Whenever values are transformed into z-scores, 95% of the data will fit somewhere between z-scores of -1.96 and 1.96. This is true no matter what range your data is. If your data has a range of 3 or 3,000,000, 95% of the data still fall between z-scores of -1.96 and 1.96. This is HUGELY helpful for determining whether problems in your data (errors) are big enough to be a big problem that throws off your analysis. So, while you won't be expected to do z-score work throughout this book, if you read far enough, you will hit a page in which we say, "Hey! Do you remember z-scores in Chapter 5?" and this little bit of information is going to be critical for understanding those chapters.

Also, look back at Figure 5.2. The standard deviations at the bottom of the curve are equal to z-scores in the normal curve (e.g., $s = 1$, $z = 1$). That's another thing that makes the normal distribution special!

Excel and SPSS

First, let's look at how to generate descriptive statistics in Excel. To generate descriptive statistics:

1. Open the Data Analysis Menu
2. Select Descriptive Statistics
3. Under input range, select the range of cells containing items in your variable
4. Under output, type the number or the cell that you want Excel to begin printing the output in
5. Make sure all boxes are checked, and the confidence level is set to the default of 95%
6. Click *OK*

Figure 5.5 shows the Descriptive Statistics window.

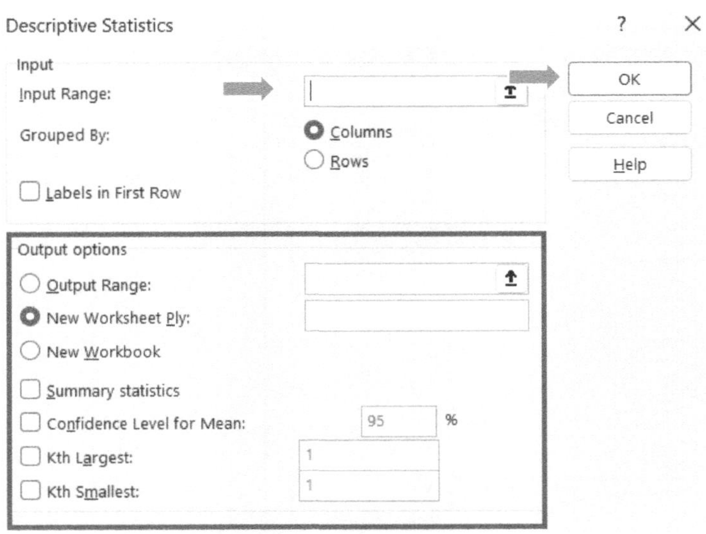

Figure 5.5 Descriptive Statistics in Excel.

To derive descriptive statistics in SPSS:

1. Open the Data Analysis menu
2. Open the Descriptive Statistics menu and choose Descriptives (see Figure 5.6 for the location of these selections)
3. Once you've selected Descriptives from the menu, a new box will open called Descriptives
4. This box will have a button called Options. . . Click on Options. . . to choose the descriptive statistics you wish to be output. We always recommend Mean, Minimum, Maximum, Kurtosis, and Skewness
5. Click Continue in the Options. . . menu. This will close the options window
6. Select the name of any variables you wish to see the descriptive statistics of from the left-hand box and click the arrow to move it to the right-hand box labeled Variable(s)
7. Click OK

The location of these buttons is shown in Figures 5.6 and 5.7.

An advantage of doing descriptive calculations in SPSS rather than Excel is that you can calculate descriptive statistics for multiple variables simultaneously. These selections can also be made faster because SPSS already knows the range of each variable, thanks to the setup you completed in the Variable View. Remember, the steps for setting up data in the Variable View are overviewed in Chapter 3, if you need a refresher.

Summary

Much like Lego Captain America in his journey to beautify Lego Land, your journey through descriptive statistics has equipped you with the tools to bring order to data that might otherwise seem chaotic or overwhelming. You've learned to use central tendencies to grasp the essence of our

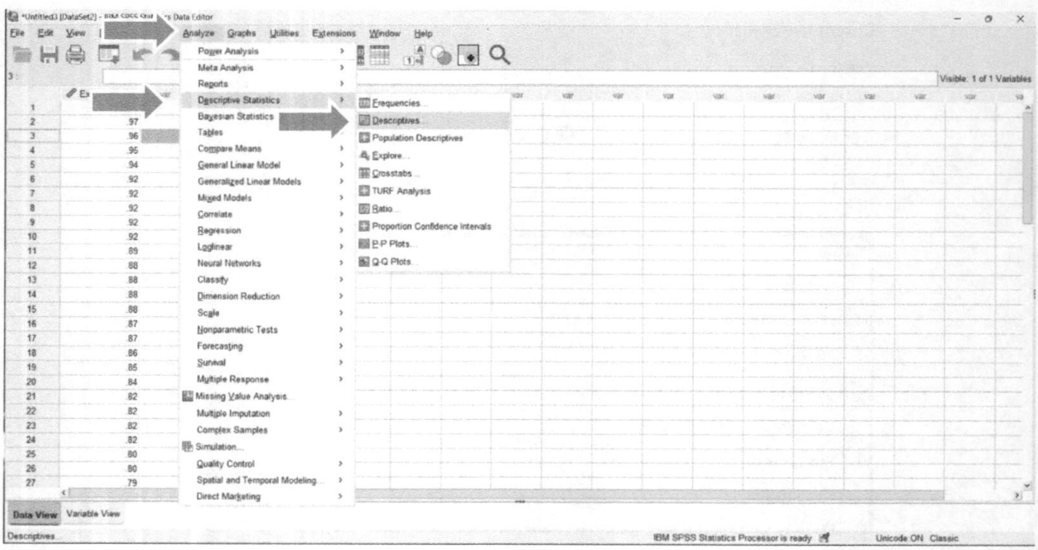

Figure 5.6 Descriptive Menu in SPSS.

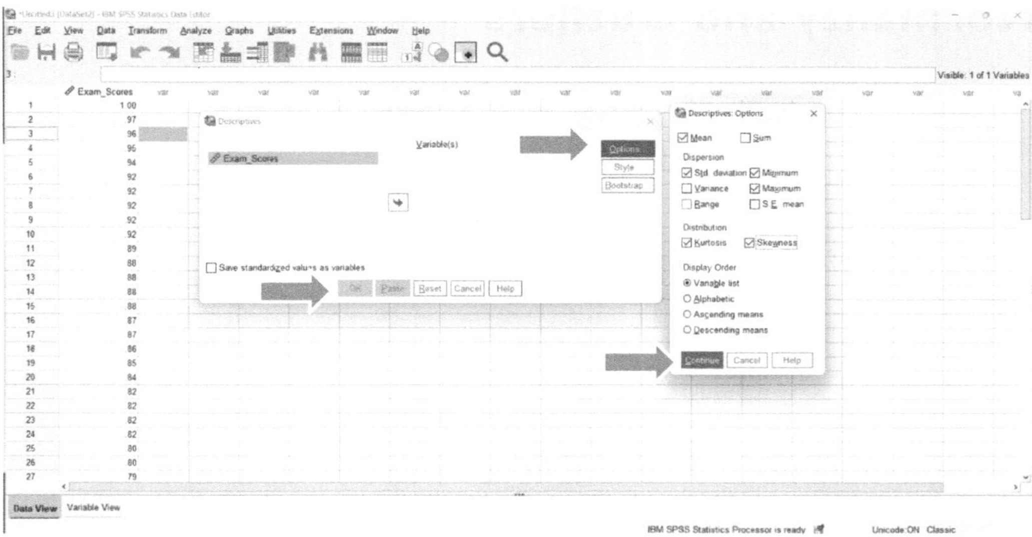

Figure 5.7 Descriptive Options in SPSS.

data, preparing you to describe samples and measures. You know how to use standard deviation and variance to imagine how spread out your data is. You can visualize your data through histograms to describe it with shapes rather than numbers. Just as Lego Captain America's understanding of his materials enabled him to craft a world, your mastery of descriptive statistics empowers you to uncover insights and build understanding from your data.

References

Croucher, S. M., Kelly, S., Burkey, M., Spencer, A., Gomez, O., Del Villar, C., & Eskiçorapçı, N. (2021). A multi-national validity analysis of the argumentativeness measure. *International Journal of Conflict Management*, *32*(1), 88–101. https://doi.org/10.1108/IJCMA-02-2020-0027

Zeng, C., Kelly, S., & Goke, R. (2020). Exploring the impacts of leader integrity and ethics on upward dissent and whistleblowing intentions. *Communication Reports*, *33*(2), 82–94. https://doi.org/10.1080/08934215.2020.1761413

Chapter 6

Normality and Outliers

What Will I Learn in This Chapter?

Steve was raised in the Empire, and so his first career choice was always going to be a Stormtrooper. His father and his brothers were all Stormtroopers, and he wanted to be just like them. He went to the Imperial Academy and tried to be just like them. He soon realized this was not a good decision. Not only did his dad die on the first Death Star, but his brothers perished on the second Death Star. In addition, he (unlike so many other Stormtroopers) realized he never hit his target with his laser. He quickly came to the realization that his family had biased his career choice. He wasn't cut out for infantry. Steve loved programming computers

DOI: 10.4324/9781003500117-6

and recognized that his family had biased his decision-making his whole life, preventing him from considering the possibility of a future working with computers. A similar thing can happen in statistics; bias can be present in the data and influence our analysis by keeping us from seeing the big picture.

In this chapter, we will learn all about bias, and many other statistical assumptions. Specifically, in this chapter, we will cover the following:

1. What is statistical bias?
2. Key assumptions of parametric data: interval level data, independence, homogeneity of variance, and normality
3. What are statistical outliers?
4. How to spot outliers in SPSS
5. Statistically define normality using SPSS
6. Assess normality and variance using SPSS
7. Methods of reducing bias

Statistical Bias

Everyone is guilty of bias. If you are a sports fan, how many times have you accused a referee of being biased against your favorite team? It's common to accuse certain media outlets of being biased against political parties and candidates. In such cases, bias represents a systematic distortion or prejudice against a person or group. In statistical terms, **bias** is a systematic distortion of a statistical result due to unaccounted-for factors. No matter how sophisticated our statistical tools are, we cannot predict or understand everything perfectly, as there is always some amount of error, or unexplained variance (we discussed this in Chapter 5). Thus, our results are estimates, which must take into consideration bias. Bias comes from two main sources: violations of statistical assumptions and outliers.

Statistical Assumptions

An assumption is a condition or term that guarantees what you are trying to do will work. For example, when driving a car, you are under the assumption that when you put pressure on the gas pedal, the car will accelerate, because putting pressure on the gas pedal should signal the engine to move the car. Granted, there are other assumptions that must be in place for this to work: the car must be turned on, there must be gas in the car, other parts of the engine must be working, etc. If any of these conditions fail, then putting your foot on the gas might not do a thing. It is the same with statistical assumptions. **Statistical assumptions** are a set of conditions that relate to the quality of our statistical procedure(s). What follows is a discussion of four statistical assumptions that are based on a normal distribution, **parametric tests**. If you refer to Chapter 2, we discussed the Central Limit Theorem and what constitutes a normal distribution and then shared more about the normal distribution in Chapter 5. Unless you have an opportunity to do experimental work with your mentors, odds are that most of the analyses you will need to work with data in your degree will be parametric. Other analyses, called "non-parametric tests," are useful, but not as commonly needed for students in communication. While parametric tests require assumptions that accompany a normal distribution, **non-parametric tests** do not, though they have their own assumptions, which will be discussed in chapters that cover non-parametric tests. The four assumptions of parametric data are: interval level data, independence, normality, and homogeneity of variance.

Test Your Understanding 6.1 – What is the difference between a parametric and a non-parametric test?

The first statistical assumption is that the data should be measured at least at the interval level. Refer to Chapter 2 for a discussion of interval level data. In brief, it is numeric data without a true zero.

The second statistical assumption is independence. **Independence** can mean different things depending on the test. Independence refers to the idea that the data points (scores) are not influenced by one another. For example, if Stormtrooper Steve and his brother trained under the same combat instructor, their skill levels (and those of their classmates) are not independent because they were influenced by the same teaching practices. However, there are scenarios where comparing Steve and his brother's classes is still valid. For instance, you might compare the effectiveness of training across different time periods or with varying selection criteria for cadets. In these cases, the groups are considered independent because the comparison is based on variables that differ between them (e.g., time or selection criteria). What you cannot do is compare outcomes specifically tied to teaching effectiveness, as these would be dependent on the instructor and, therefore, violate the assumption of independence. The third statistical assumption is **normality**. A key to hypothesis testing, and research, is that data is normally distributed. As the Central Limit Theorem posits, under normal conditions, the distribution of a sample mean converges to the mean of a population and data from larger samples tends to be more normally distributed. In addition, when more samples are taken from a population, our chances of accurately representing that population increase. With normally distributed data, we are more accurately able to generalize our findings from a sample to our intended population. Later in this chapter, we will illustrate how to describe normality using SPSS.

There are multiple ways to describe and test for normality in a dataset. One of the first ways to test for normality is to examine the skewness and kurtosis of the data. These terms were defined, and you were shown how to generate these statistics in Chapter 5. To review, skewness describes how symmetrical your data distribution is, while kurtosis describes the flatness or pointiness of your distribution. The closer your skewness and kurtosis scores are to zero, the more normally distributed your data is. Data can be considered "normally" distributed if skewness levels range between −2 and +2 and kurtosis levels range between −2 and +2.

Test Your Understanding 6.3 – What is the difference between skewness and kurtosis?

Aside from assessing the skewness and kurtosis of a distribution, you can examine if your distribution deviates from a normal distribution using the Kolmogorov–Smirnov and Shapiro–Wilk tests. Both tests compare the scores in your dataset to a normally distributed dataset with the same mean and standard deviation. If the test is non-significant ($p > .05$), the distribution of your sample compared to the normal distribution is not significantly different. If the test is significant ($p < .05$), this means your data is significantly different from a normal distribution (it violates normality). However, both tests have a significant limitation; they are impacted by the size of your sample. In large samples, even small deviations from normality can yield statistically significant results, making it important to evaluate the practical impact of these deviations. Use visualizations and other descriptive measures, such as skewness and kurtosis, to assess whether the deviations are meaningful in the context of your research. While in smaller samples (where you should worry more about normality), the test may not be able to fully detect differences. The Shapiro–Wilk test is more appropriate for small sample sizes ($n < 50$), while the Kolmogorov–Smirnov test is best for larger sample sizes ($n > 50$). Note, though, that the Kolmogorov–Smirnov test becomes unreliable with samples of 100 or larger as the power of the sample interferes with the output. For larger samples, other methods of determining normality are needed. The key is to not only run these tests but also look at skewness and kurtosis, and to graph your data and look for patterns. (See Chapter 4 for a review on making charts.) Looking at the graphical representation of your data is particularly important for small sample sizes because a non-significant Shapiro–Wilk test may be an issue of sample size rather than non-normality.

Test Your Understanding 6.4 – What impact does sample size have on the Kolmogorov–Smirnov and the Shapiro–Wilk tests?

You can also visually assess the distribution of your data using methods likely familiar to you since middle school, or at least since Chapters 4 and 5. Recall that a histogram is a visual plot of the frequency of each score in your dataset. The histogram shown in Figure 6.1 depicts the distribution of a dataset. As you can see, this figure is almost normally distributed (i.e., almost has a perfect bell curve shape), but is somewhat left skewed because of a low score to the left. Take note of that low score. We'll be talking more about what to do when you see something like this in your data later in the chapter.

Another method for looking at distributions is the box and whiskers plot. Figure 6.2 shows a box and whisker plot of the same data depicted in Figure 6.1. A perfect normal distribution has boxes and whiskers of the same length with all data fitting onto the box and whiskers.

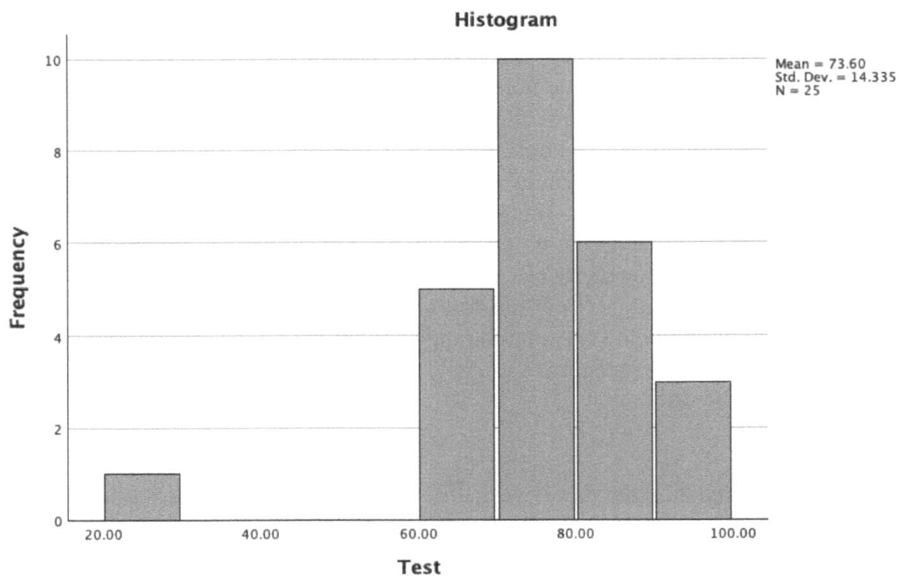

Figure 6.1 Histogram.

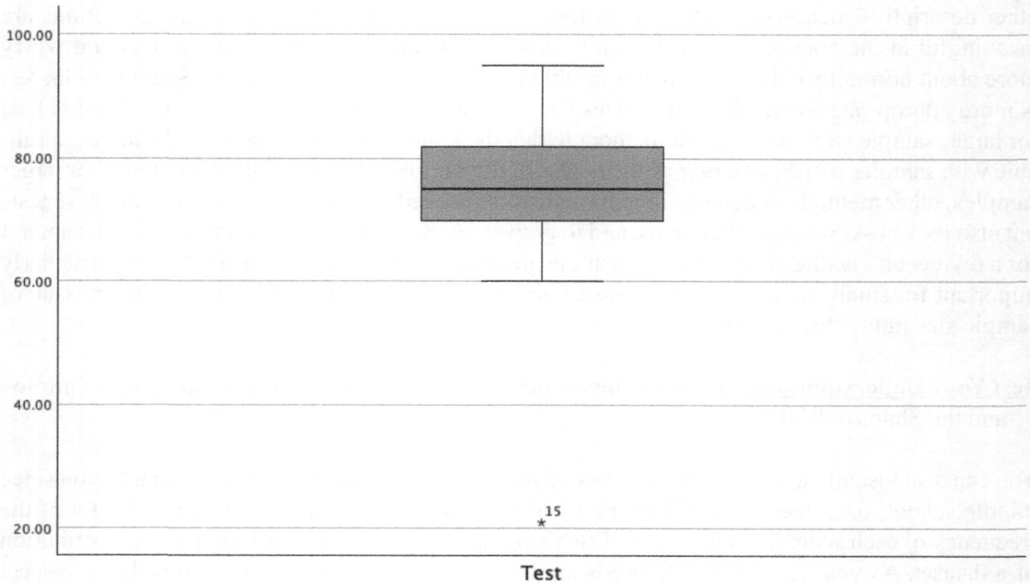

Figure 6.2 Boxplot.

While the boxes and whiskers are almost the same length in this box and whisker plot, there is a single datapoint, that lone low score, floating below the second whisker, which throws off the distribution.

It's wise to look at either a histogram or a box and whisker plot before running any analyses on your data. As discussed in Chapter 4, humans are visual creatures. You'll see peculiarities in your data much faster in these formats than you will from a table of outputs.

The fourth statistical assumption is homogeneity of variance. **Homogeneity of variance** is the statistical assumption that variances should be similar throughout the data. This assumption is discussed at length throughout this book, and later in this chapter, we will illustrate how to test for homogeneity of variance. Homogeneity of variance is the statistical assumption that variances should be similar throughout the data. This means that if you have multiple groups, the dispersion of your data should look similar across groups if you want to compare those groups. Levene's test is used to statistically test homogeneity of variance. **Levene's test** tests a null hypothesis that variances in different groups are equal. Levene's test also assumes homoscedasticity, which refers to the spread of data above and below a common range, but let's table the discussion of homogeneity until we've learned a bit more statistics.

Outliers

An **outlier** is a score or data point that is very different from the other scores or data points, such that it is more than two standard deviations removed from the mean of your data. Outliers can significantly impact measures of central tendency, and measures of variability. Let's look at the dataset with the outlier that you saw in Figures 6.1 and 6.2. Imagine that the data in Table 6.1 is from a class's statistics test:

Table 6.1 Scores on the Statistics Test

85	84	78	90	69
82	65	90	60	81
70	21	60	80	70
71	75	95	85	70
70	75	72	65	77

The outlier is 21.

This is the score that stood out in both the histogram and the box and whiskers plot. This score is much lower than all the others. The mean (average) score on the test was 73.60, with a standard deviation of 14.34 and a variance of 205.50. Two standard deviations below the mean is 44.92. Thus, a score of 21 is more than two standard deviations below the mean and is considered an outlier. In fact, it's more than three standard deviations below the mean! Thinking back to Chapter 5, we know that if something is more than three standard deviations removed from the mean, then there is less than a 1% chance of that score appearing in the population. This points to the real problem with outliers – they are often signs of **sampling error**, an instance in which a piece of data accidentally appeared in our sample that doesn't belong.

Consider how a student scores a 21 on a test. This score is so different than the others that if the 21 was removed from the dataset, the mean becomes 75.79, the standard deviation = 9.44, and the variance = 89.13. This outlier significantly influences the variability of the data. So, does this student's score belong in the sample as we assess how the class performed? There is a high probability that if the purpose of the exam was to assess what students have learned from the course material, this student was tested before learning occurred. Was this a student who was absent when the content was covered and had the misfortune of returning to class on exam day before excuse documentation was verified? Is this a student who has no intention of learning the material, only registered in the class to be full-time, and only shows up on exam days for attendance verification? Something is different about this student; they are not like everyone else in the class.

If you were trying to understand how well the material was conveyed to students, this one score would greatly impact how you interpret the variability of the results. The more bias there is due to outliers in this equation, the less confidence we can have in our results. This is why statisticians have given an immense amount of consideration to the problem of dealing with outliers.

<u>**Test Your Understanding 6.4**</u> – What impact(s) can outliers have on a dataset?

Spotting Outliers in SPSS

Before we go further, let's look at other methods of spotting outliers that may not be familiar to you.

1. Go to Data
2. Select Identify Unusual Cases
3. Move the variable you want to analyze over to Analysis Variables
4. Click OK

Figure 6.3 shows you how to pull up unusual cases in SPSS.

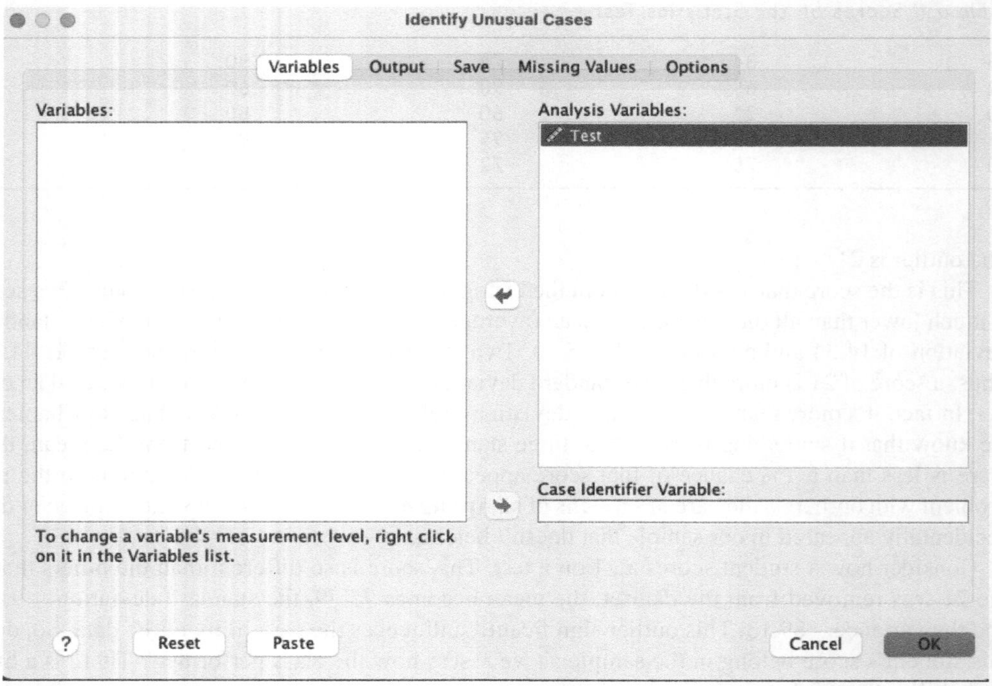

Figure 6.3 How to identify unusual cases.

Table 6.2 Anomaly Case Reason List

Reason: 1				
Case	*Reason Variable*	*Variable Impact*	*Variable Value*	*Variable Norm*
15	Test	1.000	21.00	21.0000

You will get three outputs (Table 6.2). The key is the third output, which gives you the value of the outlier and the case number in SPSS, which appears in a table called "Anomaly Case Reason List" and is shown in Table 6.2.

The second way to identify outliers is to explore the data visually.

1. Go to Analyze
2. Click on Explore
3. Move your variable over to the Dependent List. This is shown in Figure 6.4
4. Click on Statistics and choose Descriptives and Outliers
5. Then click Continue. See Figure 6.5.
6. Next, choose Plots and click Stem-and-leaf and Histogram
7. Click Continue. Once back to the main Explore page
8. Click Continue. See Figure 6.6.

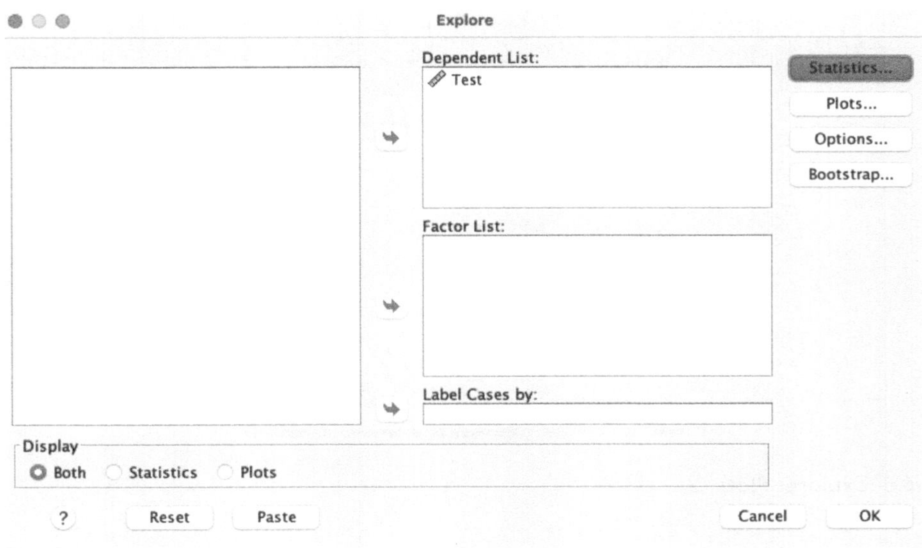

Figure 6.4 Explore data.

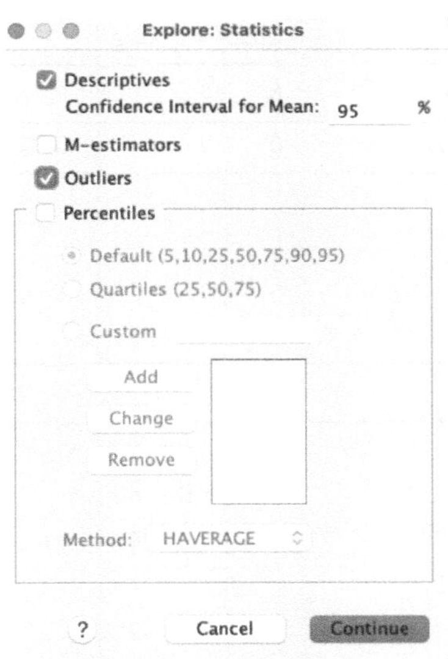

Figure 6.5 Explore: Statistics.

You will get five relevant outputs from this analysis. You've seen two of them – the histogram and box and whisker plot in Figures 6.1 and 6.2. The next output is a summary. As shown in Table 6.3, this summary provides the sample size, including an overview of any missing data.

The second new output provides all the descriptive statistics for your data, in this case, the statistics test, as shown in Table 6.4. Here you get the mean, standard deviation, skewness, and

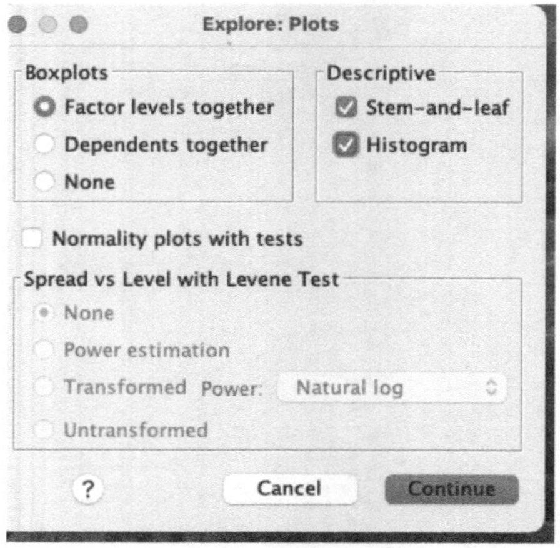

Figure 6.6 Explore: Plots.

Table 6.3 Case Processing Summary

	Valid		*Cases Missing*		*Total*	
	N	*Percent*	*N*	*Percent*	*N*	*Percent*
Test	25	100.0%	0	0.0%	25	100.0%

Table 6.4 Descriptives

			Statistic	*Std. Error*
Test	Mean		73.6000	2.86705
	95% Confidence Interval for Mean	Lower Bound	67.6827	
		Upper Bound	79.5173	
	5% Trimmed Mean		74.9556	
	Median		75.0000	
	Variance		205.500	
	Std. Deviation		14.33527	
	Minimum		21.00	
	Maximum		95.00	
	Range		74.00	
	Interquartile Range		13.50	
	Skewness		−1.968	.464
	Kurtosis		6.909	.902

Table 6.5 Extreme Values

			Case Number	*Value*
Test	Highest	1	25	95.00
		2	13	90.00
		3	14	90.00
		4	11	85.00
		5	12	85.00
	Lowest	1	15	21.00
		2	2	60.00
		3	1	60.00
		4	4	65.00
		5	3	65.00

kurtosis information that you need to help determine whether the data is normally distributed and whether a score far from the mean is, in fact, an outlier.

The third output (Table 6.5) shows the extreme values in the data.

The fourth output is a histogram that shows you the distribution of the data. In this visualization, you can see that there is one case that stands out as distinct from all the rest. Combined, these techniques identify outliers in your data. Later in this chapter, we will discuss how to manage or what to do with them.

Test Your Understanding 6.6 – What are two procedures to identify outliers in data?

The last normality exploration to cover in SPSS is the actual tests for normality, not just the inferences we've looked at so far. Using SPSS to test for the Kolmogorov–Smirnov and Shapiro–Wilk tests is accomplished by running the same commands in SPSS. Let's explore this with some data on people's fear of COVID-19 in the United Kingdom. A total of 367 people participated in the project ($n = 367$). To explore normality on this dataset, the following steps are followed in SPSS.

1. Click Analyze
2. Choose Descriptive Statistics
3. Choose Explore (Review Figure 6.6 if you need assistance finding the Analyze menu)
4. Then select the variable you want and move it over to the Dependent List. In this case, we are choosing *PUBLHEALTH*, which explains people's belief that COVID-19 is a threat to public health (Figure 6.7)
5. The next step is to choose Plots and pick Normality plots with tests
6. Click Continue. When you are back to the main Explore page, click *OK* (Figure 6.8)

After you click OK, you will get multiple outputs. The following are key. The first output, Table 6.6, summarizes the data. It shows how many people completed this measure (*PUBLHEALTH*), and if there was any missing data. As shown in Table 6.6, only 68.4% of the participants completed this measure.

The second output is the Descriptives. Table 6.7 provides all the descriptive statistics. When assessing normality, you will notice that kurtosis and skewness are reported in this output. Both fall within the ranges discussed earlier for normally distributed data.

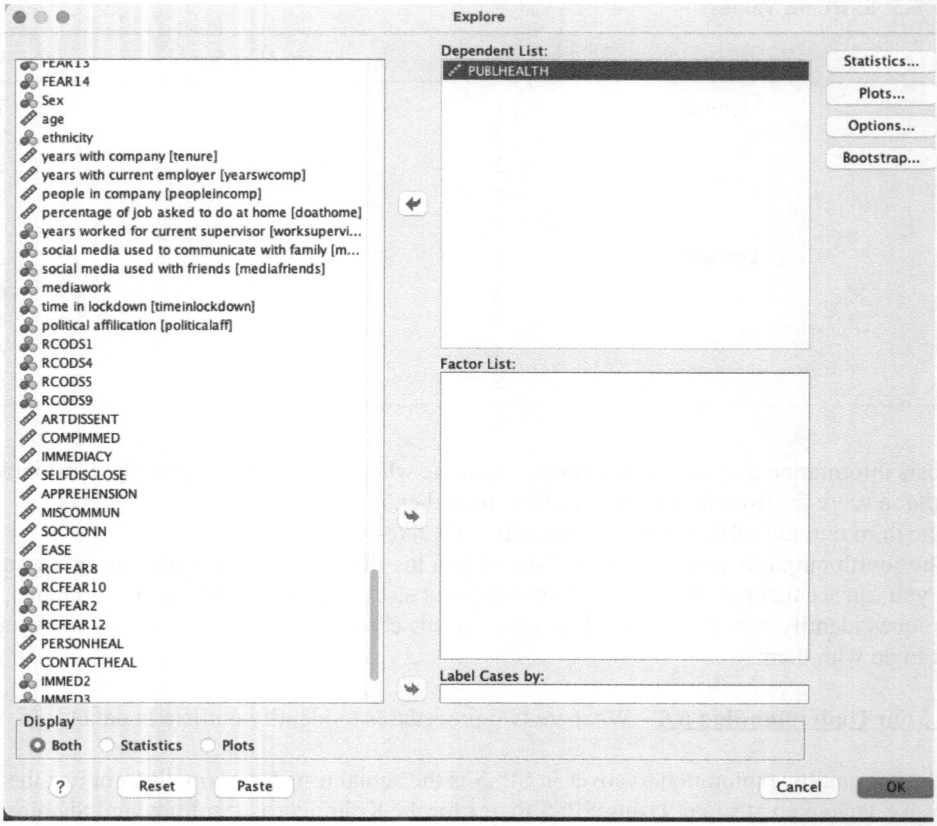

Figure 6.7 Explore normality.

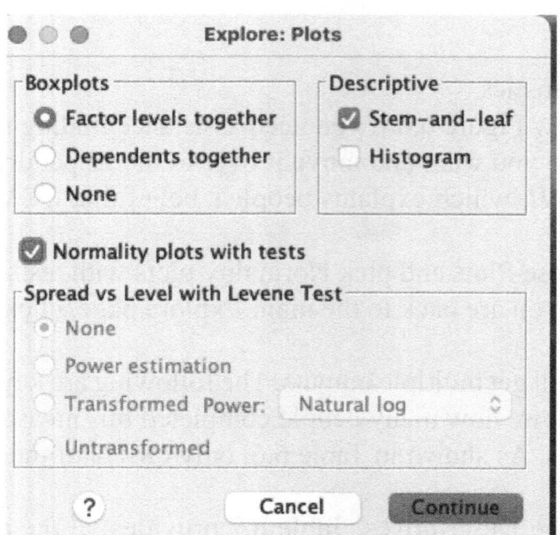

Figure 6.8 Explore normality plots.

Table 6.6 Case Processing Summary

	Valid		*Cases Missing*		*Total*	
	N	*Percent*	*N*	*Percent*	*N*	*Percent*
PUBLHEALTH	251	68.4%	116	31.6%	367	100.0%

Table 6.7 Descriptives

			Statistic	*Std. Error*
PUBLHEALTH	Mean		1.0817	.05098
	95% Confidence Interval for Mean	Lower Bound	.9813	
		Upper Bound	1.1821	
	5% Trimmed Mean		1.0431	
	Median		1.0000	
	Variance		.652	
	Std. Deviation		.80765	
	Minimum		.00	
	Maximum		3.00	
	Range		3.00	
	Interquartile Range		1.00	
	Skewness		.433	.154
	Kurtosis		−.396	.306

Table 6.8 Tests of Normality

	Kolmogorov–Smirnov[a]			*Shapiro–Wilk*		
	Statistic	*df*	*Sig.*	*Statistic*	*df*	*Sig.*
PUBLHEALTH	.178	2S1	<.001	.919	2S1	<.001

a. Lilliefors Significance Correction

The third output, Table 6.8, is the Tests of Normality. This output reports the results of the Kolmogorov–Smirnov and Shapiro–Wilk tests. Both tests are significant, which means the data is significantly different from a normal distribution. Keep in mind, this is a large sample. The result of both tests is impacted by the sample size. Therefore, when considering the normality of this data, we need to look at these two tests, skewness, kurtosis, and possible outliers. The authors of your book, knowing the sample size, would rely on the low skewness and kurtosis results to state that the data in this case meets the assumption of normality.

Homogeneity of Variance Using SPSS

Running Levene's test in SPSS is simple. Just like running the Kolmogorov–Smirnov and Shapiro–Wilk tests, the following steps are followed in SPSS.

1. Click Analyze
2. Choose Descriptive Statistics
3. Choose Explore
4. Then select the variable you want and move it over to the *Dependent List*. In this case, we are choosing *CovidPubHealth*, which explains people's belief that COVID-19 is a threat to public health
5. Next, you need to choose the factor or group that you want to examine. In this case, we will be examining the variance in educational level on *CovidPubHealth*. Therefore, choose Education and move it over to the Factor List (Figure 6.9)
6. The next step is to choose Plots and pick Factor levels together, Stem and leaf, Histogram, Normality plot with tests, Transformed, and Natural log
7. Then press Continue
8. When you are back to the main Explore page, click OK (Figure 6.10)

After you click OK, you will get multiple outputs. The following three are key. The first output, Table 6.9, summarizes the data. It shows how many people completed this measure and if there are any missing data points.

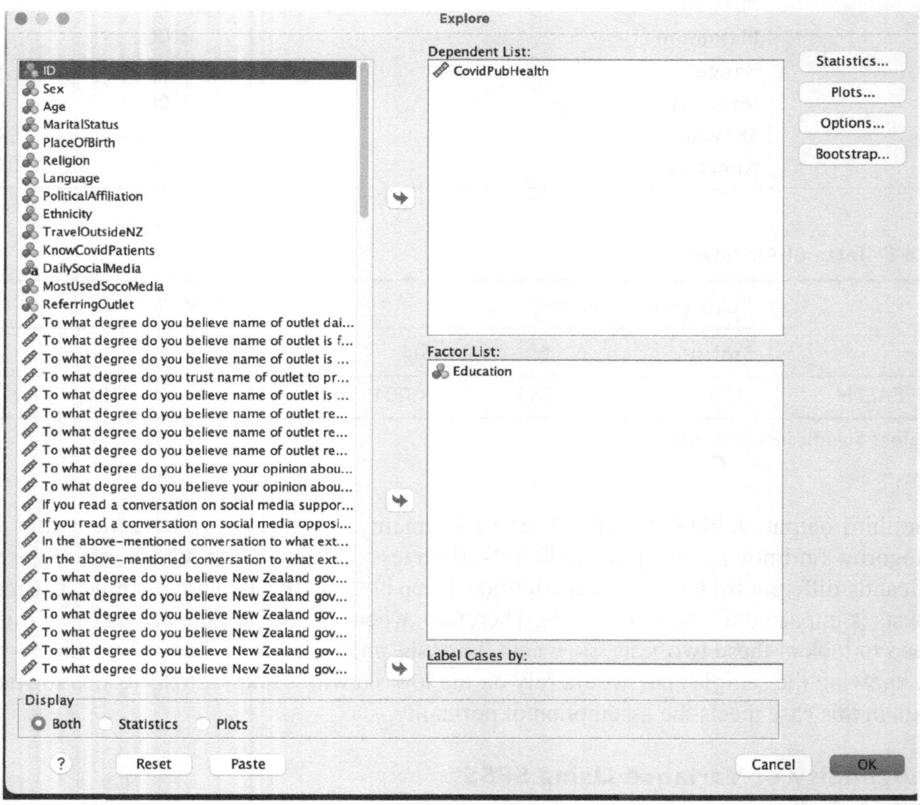

Figure 6.9 Explore Levene's test.

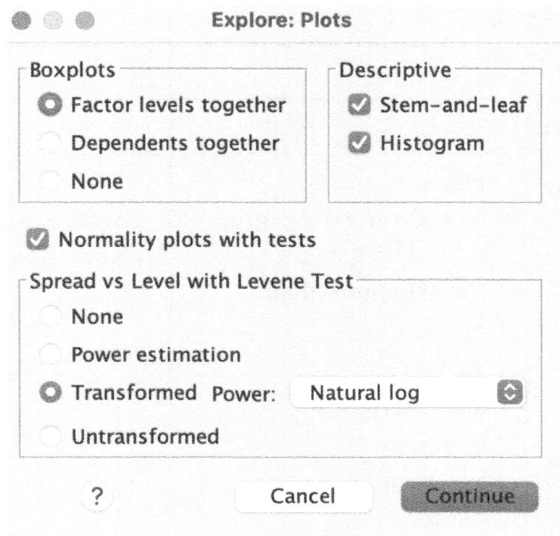

Figure 6.10 Explore normality plots – Levene's.

Table 6.9 Case Processing Summary – Levene's Test

		Valid		*Cases Missing*		*Total*	
	Education	*N*	*Percent*	*N*	*Percent*	*N*	*Percent*
CovidPubHealth	High school	137	98.6%	2	1.4%	139	100.0%
	2-year degree	43	97.7%	1	2.3%	44	100.0%
	4-year degree	85	100.0%	0	0.0%	85	100.0%
	Masters	33	100.0%	0	0.0%	33	100.0%
	Doctorate	6	100.0%	0	0.0%	6	100.0%

The second output is the Descriptives. Table 6.10 provides all the descriptive statistics for *CovidPubHealth* divided up by each educational group.

The *final* key output for Levene's test is the Test of Homogeneity of Variance Output. Table 6.11 shows whether Levene's is significant. In this case, Levene's is not significant, which means the variances are not significantly different and we have homogeneity.

If you *wanted* to report the results of Levene's test, you would report the results as such: the variances between educational levels on public health fears of COVID-19 were not significantly different, $F(4, 299) = .47, p = .76$. Note that the 4 and 299 come from the degrees of freedom based on means (number of values that can change in a calculation without breaking any rules), the .47 is the F-value, and .76 is the significance level.

Ways to Reduce Bias

Throughout this chapter, we have talked about bias and how distortions and/or errors in our data impact our results. Outliers, normality issues, and issues with homogeneity of variance are three big culprits of bias. So, what can we do to reduce bias? There are three things we

Table 6.10 Descriptives – Levene's Test

Descriptives

	Education			*Statistic*	*Std. Error*
CovidPubHealth	High school	Mean		2.7981	.06094
		95% Confidence Interval for Mean	Lower Bound	2.6775	
			Upper Bound	2.9186	
		5% Trimmed Mean		2.7833	
		Median		2.6667	
		Variance		.509	
		Std. Deviation		.71327	
		Minimum		1.00	
		Maximum		S.00	
		Range		4.00	
		Interquartile Range		1.00	
		Skewness		.484	.207
		Kurtosis		1.492	.411
	2-year degree	Mean		2.8372	.10369
		95% Confidence Interval for Mean	Lower Bound	2.6280	
			Upper Bound	3.0465	
		5% Trimmed Mean		2.8105	
		Median		2.6667	
		Variance		.462	
		Std. Deviation		.67992	
		Minimum		1.67	
		Maximum		4.67	
		Range		3.00	
		Interquartile Range		1.00	
		Skewness		.618	.361
		Kurtosis		−.150	.709
	4-year degree	Mean		2.7569	.06932
		95% Confidence Interval for Mean	Lower Bound	2.6190	
			Upper Bound	2.8947	
		5% Trimmed Mean		2.7342	
		Median		2.6667	
		Variance		.408	
		Std. Deviation		.63909	
		Minimum		1.33	
		Maximum		5.00	
		Range		3.67	
		Interquartile Range		1.00	
		Skewness		.693	.261
		Kurtosis		1.134	.SI 7

(Continued)

Table 6.10 (Continued)

Descriptives

	Education			*Statistic*	*Std. Error*
	Masters	Mean		2.7879	.09492
		95% Confidence Interval for Mean	Lower 8ound	2.5945	
			Upper Bound	2.9812	
		5% Trimmed Mean		2.7716	
		Median		2.6667	
		Variance		.297	
		Std. Deviation		.54530	
		Minimum		2.00	
		Maximum		4.00	
		Range		2.00	
		Interquartile Range		.83	
		Skewness		.144	.409
		Kurtosis		−.619	.798
	Doctorate	Mean		3.3333	.38490
		95% Confidence Interval for Mean	Lower Bound	2.3439	
			Upper Bound	4.3228	
		5% Trimmed Mean		3.3148	
		Median		3.3333	
		Variance		.889	
		Std. Deviation		.94281	
		Minimum		2.33	
		Maximum		4.67	
		Range		2.33	
		Interquartile Range		1.83	
		Skewness		.239	.845
		Kurtosis		−1.481	1.741

Table 6.11 Test of Homogeneity of Variance

		Levene Statistic	*df1*	*df2*	*Sig.*
CovidPubHealth	Based on Mean	.467	4	299	.760
	Based on Median	.429	4	299	.788
	Based on Median and with adjusted df	.429	4	266.950	.788
	Based on trimmed mean	.493	4	299	.741

can do to reduce and/or mitigate the effects of bias. First, be sure your study is methodologically sound. Before you even collect data, make sure you have considered issues of independence, measurement level, sampling, reliability, validity, etc. If you take steps to ensure your project is methodologically and theoretically sound, you can reduce the chances of bias popping up later. However, when you explore your data, you will more than likely find outliers, issues with normality, and variance issues. There are ways to manage these practically and statistically.

Ways to Manage Outliers

There are many options you can use to manage the impact of outliers in your data. You first need to explore the data and see if you have any outliers. If you do, then you can use one of these options below. The key is to explain to the reader what you did to manage the outliers when you are writing up your research report, thesis, or article.

1. Remove the outliers – You delete the outliers from the data. For example, let's say you have 250 participants, and you identify from your exploration that participant numbers 10, 151, and 219 are outliers. You delete them from the analysis. Re-run the data and see if the removal of these individual participants changes the mean, standard deviation, and other descriptives. In smaller datasets, this often will solve the issue. For full transparency, be sure you report that you deleted *X* number of participants who were outliers.
2. Transform the data – As discussed earlier, outliers often skew data. Skew can sometimes be minimized by transforming data. This will be discussed in the next section under how to manage non-normality and unequal variances.
3. Change the Score – A third option is to adjust the value of the outlier. While this might seem like data "cheating" or manipulation, it can be appropriate when the outlier is clearly unrepresentative (e.g., due to data entry errors). If this option is chosen, researchers must transparently report such adjustments, justify their reasoning, and discuss potential impacts on the analysis.
    a. Replace the outlier(s) with the mean of the variable. This is the one most often used by the authors of this book when there are few outliers.
    b. Replace the outlier(s) with the median of the variable. This is the option used by the authors of this book when there are a significant number of outliers, and the mean and median are significantly different from one another.

This may still sound like cheating, so let's consider an example of when replacing the outlier may be a reasonable choice. Imagine that you were analyzing student data and found someone listed as 221 years old. Having a college student over 200 years old in the dataset is going to skew your distribution. Logically, we know that is not the correct age and can assume that likely this person was either 22 or 21, but held their finger on one of the keys too long when typing their age. We do not know which is right, so we do not want to assign them 22 or 21, which are equally likely to be wrong. Instead, we can delete 221 and replace age with the mean or median age, which will not change the shape of our distribution beyond correcting skew, but is our best guess of the true age of a participant in that sample.

Ways to Manage Non-Normality and Unequal Variances

When you encounter issues with non-normality and unequal variances, a technique you can apply is to statistically transform the data. Just like replacing data sounded like cheating, transforming data might sound a bit sketchy. However, the argument behind transformations is that you alter every score or data point equally to address outlier, distributional, or unequal variance issues. While transformations do not alter the relationships or differences between variables, they can complicate interpretation. For example, results may need to be back-transformed to their original scale, which can introduce additional complexity when reporting and discussing findings. There are four primary transformations you can choose from, and your choice of transformation depends on the type of skew in the data and the data's range. In this book, we will not go into depth on how to compute these transformations. You can look to more advanced statistical books, such as Field (2009), for tutorials on how to compute the following three transformations.

1. Logarithmic transformation – When a dataset is positively skewed and has unequal variances, you can take the logarithm of the data to minimize the skew. As this transformation is based on logarithms (think back to algebra), you can't have a log value of 0 or negative numbers. So, if your dataset has negative numbers or the means are close to 0, you will need to add a constant to the data before performing the transformation. Notably, this transformation also reduces heteroscedasticity.
2. Square root transformation – When a dataset is positively skewed and has unequal variances, you can also take the square root of every score. This procedure will pull higher scores closer to the center of the distribution. You will still have the same mathematical problem with negative numbers, though. Like the log transformation, the square root transformation will reduce heteroscedasticity.
3. Exponential transformation – Exponential transformations are often applied when data is negatively skewed or to deal with unequal variances across groups. This transformation works by raising a base (your variable) to a power. Unlike logarithmic and square root transformations, exponential transformations stretch the scale of the data rather than compress it, which can make smaller differences in the lower range of the data more pronounced. However, this approach should be used cautiously, as it can exacerbate heteroscedasticity if applied inappropriately. Additionally, any negative or zero values in the dataset will need to be adjusted by adding a constant. For example, if your interval scale ranges from −3 to 3, you'll need to add 4 to all variable values before applying the transformation. Otherwise, −3 and 3 will have the same value if you raise them to an even power. As with other transformations, be sure to document and explain their application thoroughly in your research to ensure clarity and replicability.
4. Reverse score transformation – Reverse score transformations are sometimes used when correcting for negative skew in the data (see Chapter 3). It's important to note that reverse scoring does not inherently correct skewness but rather adjusts the direction of the skew to make it positive, allowing for the application of transformations that work effectively on positively skewed data. This process requires careful documentation in your research report to ensure transparency and reproducibility. Rarely in class will you have a need to run transformations. However, you may read about such analyses in journals or may need to do them with your research.

Completing the square root and logarithmic transformation in SPSS is quite simple once you've determined which type of transformation you need to use. In SPSS, click on:

1. Transform
2. Compute Variable

At this stage, you'll be asked to give your transformed variable a new name. The authors of this book like to be straightforward, so they often call the variable whatever the original variable was named, followed by "_transformed." Keep in mind that SPSS cannot handle spaces. Then after the equal sign provided by SPSS, you need to simply type in the formula for your transformed variable. In the example of data used earlier in this chapter, the variable of interest was called test. So, to create a logarithmic transformation, the formula would read:

Test_Transformed = ln(Test)

where ln is SPSS's shorthand for taking the natural log of that value. To create a square root transformation, the formula would read:

Test_Transformed = SQRT(Test)

where SQRT is SPSS's shorthand for taking the square root of a variable. Lastly, to create an exponential transformation, use ** to indicate raising your variable to a power. Below is the formula for raising the test data to the power of 2.

Test_Transformed = Test**2

Summary

Steve the Stormtrooper has been surrounded by bias his whole life. His whole family has been Stormtroopers, and he was expected to follow in their footsteps. After exploring the assumptions in his life, varying a bit, and transforming some, he realized he was still an outlier. Upon reflection, that's ok for him.

Looking at communication research, in a recent piece published by Goke and Violanti (2025), they asserted that we may know less than we think about student-to-student and student-to-instructor communication, because much research in this field has run analyses on data without testing for the assumptions of parametric tests. This means there are a lot of statistically significant results out there about the relationships between various types of student-to-student and student-to-professor communication that could be false just because those assumptions were not checked. Making decisions about how the world works based on false information, things that seemed true because fundamental data assumptions were missed, sets us up for poor communication decisions in the future.

References

Field, A. (2009). *Discovering statistics using SPSS*. Sage.

Goke, R., & Violanti, M. (2025). Student-to-student communication: Examining and extending our conceptualization and measurement of instructional dissent. *Review of Communication, 25*(1), 49–61. https://doi.org/10.1080/15358593.2024.2395278

Chapter 7

The Research Process

What Will I Learn in This Chapter?

In 1926, when A. A. Milne published *Winnie the Pooh*, the author certainly was not trying to write a statistics lesson. However, this book popularized the shorthand "TTFN!" as Tigger often said to Pooh, meaning "ta ta for now." Almost a century later, statistics students say *TTFN* to help themselves remember the difference between a Type I and Type II statistical error, where TTFN to them means Type Two: False Negative.

We do not live in a perfect world, and researchers are only humans. That means the research process is bound to be messy and prone to errors. Yet, we do want to standardize the research process as much as we can to ensure the reliability and consistency of research results over time. This includes categorizing the errors researchers can make when testing out their research ideas. In this chapter, we will delve into the fundamentals of the research process by going through:

1) What is a hypothesis?
2) Hypothesis writing and testing

DOI: 10.4324/9781003500117-7

3) Significance testing
4) Type I & II errors

Hypothesis Testing

What Is a Hypothesis?

One of the hallmarks distinguishing scientific inquiries from fake or pseudoscience is **falsifiability**. A hypothesis is falsifiable if it is possible to test whether it is false. We cannot test the hypothesis, for example, that ghosts do not exist. We would have to be able to observe everywhere all at once to ascertain that there are no ghosts anywhere. In science, we are dealing with things that can be directly or indirectly observed. Thus, our predictions are falsifiable.

A testable hypothesis is the foundation of scientific inquiry. In social sciences, a **hypothesis** refers to a tentative statement or assumption about the relationship between two or more variables. It is essentially an educated guess about the expected relationship or differences between variables. Hypothesizing starts with a research interest. After reviewing relevant studies, researchers then develop the study interest into a testable statement. For instance, you may have noticed that students who sit in the front of the classroom tend to have higher engagement, and thus are more likely to perform better on class assignments and exams. Then, you can hypothesize: students who sit in the front rows compared to those who sit in the back have higher grade point averages (GPAs) for the class.

Most quantitative research articles have one or more hypotheses in their studies. Keep in mind that a hypothesis is, after all, a guess. That means it does not have to always be right. In fact, it is commonplace that a hypothesis is rejected, or the exact opposite is supported in a study. These unexpected results can also be valuable, as researchers might learn more about the subject with non-significant or contradictory findings.

Test Your Understanding 7.1 – Can a hypothesis exist in a question form?

How Hypothesis Testing Works

As stated in the previous section, falsifiability is a key feature of scientific inquiries. You may wonder why we are trying to falsify instead of directly proving something is true. Think about the process of judging whether someone is guilty in a criminal trial. First, we would assume the defendant is innocent until we have evidence that suggests otherwise. This legal principle, known as the presumption of innocence, is highly relevant in hypothesis testing. Now, how do we know whether there is a relationship between A and B? First, we would presume no relationship exists between A and B. When our data reveals that when A increases, B also increases, we would reject the presumption of no relationship and conclude that A and B are positively related. This process of testing the falsifiability of the presumption of no relationship is hypothesis testing. The presumption of no relationship in this example is known as the **null hypothesis (H_0)**. A null hypothesis states that no relationship or difference exists between variables and is assumed to be true until support for the opposite is demonstrated. Unlike the null hypothesis, a hypothesis developed by researchers that states how the variables are related or are different is called an **alternative hypothesis (H_1)**. You can tell we are essentially evaluating the plausibility of the null hypothesis in hypothesis testing. This process is also known as **null hypothesis significance testing (NHST).** It is important to note that null hypotheses are implied but not explicitly stated in most research papers. Researchers tend to only mention alternative hypotheses in their reports that indicate differences or relationships between the studied variables.

One-Tailed versus Two-Tailed Hypotheses

A hypothesis can be one-tailed or two-tailed, depending on whether the researchers want to test the difference/relationship in one or both directions. In other words, **one-tailed hypotheses** specify a direction, whereas **two-tailed hypotheses** do not. This is why a two-tailed hypothesis is also called a non-directional hypothesis. If you're confused, don't stress. This is an abstract idea for most students until they see an example. So, say you usually get up at 8 am and are now interested in changing your morning routine by trying to get up two hours earlier than usual. If you are certain that such a change will make you feel more energetic based on your past experience you can put forth a one-tailed hypothesis: Getting up at 6 am compared to 8 am leads to higher energy levels. The one-tail here is that we specified a positive change. However, we know getting up earlier does not work for everyone and, in some cases, may even have detrimental effects on some. If you are unsure of the effects of getting up earlier and want to allow more freedom in the hypothesis, you can propose a two-tailed one without specifying a direction: Getting up at 6 am compared to 8 am will have an impact on one's energy levels. This is a two-tailed test because our prediction is just a change, not a specified positive or negative change.

In the previous example, the rationale for using one-tailed or two-tailed hypothesis is dependent on the extent to which you are confident about the direction of the tested difference. This decision also has implications on the statistical power of the test as well as potential errors. When using a two-tailed test with a significance level of 0.05, we are allotting half of the alpha that is 0.025 to testing the statistical significance in both directions. On the other hand, a one-tailed test with a significance level of 0.05 uses all the alpha to test the statistical significance only in the specified direction. This naturally renders one-tailed tests more powerful in detecting an effect and more prone to generating false positive results (also known as Type I errors explained later in this chapter). Thus, it is important to refer to previous literature on the study subject when pondering on whether to use a one-tailed or two-tailed hypothesis.

It is not uncommon that quantitative scholars may use research questions in the research process. This usually happens when researchers have little previous knowledge about the research phenomenon and try to avoid confirmation biases. You will find different ways of forming hypotheses and research questions about the same research interest in Table 7.1.

<u>Test Your Understanding 7.2</u> – Should I use one-tailed or two-tailed hypothesis when I am not sure about the relationship between the variables?

Composing Hypotheses

Writing a great research hypothesis is crucial as it forms the foundation for your study and guides your investigation. Here are a few important elements to bear in mind when drafting research hypotheses:

1. A hypothesis is a declarative statement. This means a hypothesis should never exist in a question form.

2. A hypothesis is not a random guess and should be based on literature. Formulating a strong hypothesis requires a thorough review of existing literature related to the research question. This will help us understand what is already known and identify gaps or areas that need further exploration.
3. Hypotheses should be succinct and clear. Use concise language to specify the variables and expected differences or relationships between them. (This is why most hypotheses are one-sentence statements.)
4. A hypothesis should be falsifiable, meaning it can be proven false if the data does not support it.

A bad example of a hypothesis can be seen in an astrology prediction that says, "something important will happen to Taurus in the near future." What exactly can be classified as "something important" and how far into the future is the "near future?" With such vague language, this statement cannot be meaningfully falsified, rendering it unscientific. A good example can be found in Kelly and Gaytan's study (2020) that explores the extent to which instructors' immediate behaviors, such as smiling and making eye contact, can have an impact on students' writing apprehension. In Table 7.1, you will find the implied null hypothesis and different ways of putting forth an alternative hypothesis or research question for the study. As the researchers know from previous literature that perceived immediacy of the instructor will likely help reduce students' writing anxiety, a one-tailed, directional hypothesis is used.

There are some commonly used terms in the hypotheses that can give us an idea of what is being tested in the study. Being able to identify these key terms is a useful skill, especially when you want to go through lots of readings in less time. By and large, researchers use hypotheses to describe relationships, causation, or differences between variables. With the information in Table 7.2, you can quickly identify what the researchers are testing in the study. You are also encouraged to use these terms to precisely frame your hypotheses.

Table 7.1 Examples of Forming Hypotheses and Research Questions

Null Hypothesis	H_0: There is no relationship between instructor perceived immediacy and student writing apprehension.
One-tailed Hypothesis	H_1: There is a negative relationship between instructor perceived immediacy and student writing apprehension.
Two-tailed Hypothesis	H_1: There is a relationship between instructor perceived immediacy and student writing apprehension.
Research Question	RQ: What is the relationship between instructor perceived immediacy and student writing apprehension?

Table 7.2 Common Terms Used in Hypotheses

Relationships	*Causation*	*Differences*
Relate/Relationship	Cause	Differ/Difference
Correlate/Correlation	Affect	Compare
Associated/Association	Effect	Vary/Variation
Connected to	Predict	Higher/Lower
Linked to	Leads to	More than/Less than

Testing Significance

When it comes to the decision of accepting or rejecting a null hypothesis, it is important to remember the confidence interval, which was introduced in Chapter 6. In most social scientific studies, researchers want to be at least 95% confident that the results did not happen by chance. That is when the p-value, the significance level computed for the statistical test, is 0.05 or less. Then we can claim the null hypothesis is rejected, and the alternative hypothesis is accepted. On the contrary, if the p-value is greater than 0.05, the null hypothesis is retained, and thus we conclude that there is no relationship or difference between the variables tested. This process of calculating the probability of the null hypothesis is called significance testing.

Type I and II Errors

Despite how much we hate errors, they unfortunately do occur in the process of hypothesis testing. During the COVID-19 pandemic, most of us had the experience of testing for the COVID-19 virus. In a perfect world, we would expect a positive test result for all who carry the virus and a negative result for those who do not. However, we live in no such world, and many mistakes, such as tests conducted inadequately, ineffective/expired test kits, and result mix-ups, can lead us to incorrect conclusions. Essentially, two types of errors may take place in this process. First, a person does not have the COVID-19 virus, but the test result comes back positive. As a result, the unfortunate COVID-19-free person might be forced to be isolated from family and friends in quarantine. Second, a person has the COVID-19 virus, but the test fails to detect it. In this case, this contagious person might feel carefree about attending social gatherings and end up spreading the virus to others.

Here, the null hypothesis is that the person has no COVID-19 virus. In the first scenario, the null hypothesis is true because this person is indeed COVID-19-free, but we falsely reject the null hypothesis by claiming this person carries the virus. This is called a **Type I error**, also known as a false positive result. In the second scenario, the null hypothesis is false since the tested person has COVID-19, and thus, we should have rejected it but failed to do so. This is when we have committed a **Type II error**, also known as a false negative (TTFN!).

Which Error Is Worse?

As in the previous example, both errors could lead to rather serious consequences, with one being a COVID-19-free person forced into quarantine and the other being a contagious person spreading the virus in public. If we really had to decide, a Type II error seems to be worse, as that one contagious person has the potential to infect a high number of individuals by attending a super-spreader event. But is a Type II error always worse? Consider the following example. A group of researchers works on the development of a new medicine. The null hypothesis would be that the new medicine has no effect on patients. A Type 1 error is when researchers falsely reject the null hypothesis and claim they have an effective drug. This might lead to ineffective drugs being produced and prescribed to patients. On the contrary, a Type II error is made when researchers falsely accept the null hypothesis by concluding that their effective medicine is not working. They would miss the opportunity of inventing a cure. In this example, a Type I error appears to have more imminent consequences, as a lot of resources might be wasted on an ineffective drug that could financially and physically hurt patients, whereas no additional harm occurs because of a Type II error. The rule of thumb is that Type I errors are usually worse in innovation-related research, and Type II errors are usually worse in safety- and prevention-related research. When working on innovations and inventions, a Type II error is more tolerable as it would inform the researchers to take a conservative approach that leads to no actions and no harm. In prevention-related research, such as screening for cancer, a false alarm (Type I error) is much better than a Type II error, where a patient is given the false consolation and they would miss the best time to fight the disease.

Naturally, we want to minimize both types of errors in all research projects. However, this idea is too good to be true, as there is often a trade-off between Type I and Type II errors. To decrease Type I errors, researchers can set a lower significance level, such as a 99% confidence interval, to become more conservative when interpreting their results. However, doing so would increase the risk of making Type II errors. On the flip side, increasing the power of a test, such as utilizing a larger sample, will help decrease Type II errors, but also increase the chance of committing Type I errors. Researchers often aim to strike a balance and decide which error they are willing to tolerate more. Table 7.3 summarizes these conclusions.

Example 7.1

On January 27, 2017, President Trump signed Executive Order 13769, which banned travel to the United States from seven predominantly Muslim countries – Iran, Iraq, Libya, Somalia, Sudan, Syria, and Yemen – and suspended the resettlement of all Syrian refugees (Trump, 2017). The rationale for issuing this Executive Order was to prevent foreign terrorists from entering the U.S.A. Now, at the Port of Entry, the security screening staff might make two types of mistakes: (1) They mistake normal travelers for terrorists, and/or (2) They fail to detect terrorists among the travelers. In this example, mistaking travelers for terrorists is a Type I error, whereas failing to detect terrorists among travelers is a Type II error. By implementing this Executive Order that bans travel to the U.S.A. from these countries, the U.S. government eliminates the chance of making Type II errors but also commits itself to making Type I errors. In other words, all citizens from these countries are now considered terrorists and thus denied entry to the U.S.A. It is not surprising that this Executive Order sparked widespread protests for constituting religious discrimination and was immediately removed when President Joe Biden stepped into office in 2021.

Confidence Intervals and Significance

When conducting statistical research, we need to be certain that our results are statistically significant before we can claim that we found something significant/meaningful. In the social sciences, researchers tend to rely on a 95% confidence interval. This means a researcher can be 95% or more certain that what they found was not by chance. The remaining 5% can be attributed to various

Table 7.3 Type I & II Errors

		The Truth	
		The tested difference/ relationship is significant	The tested difference/ relationship is non-significant
Your Conclusion	I have a significant finding		Type I Error (False positives)
	I don't have a significant finding	Type II Error (False negatives)	

- $p > .05$, there is more than a 5% chance there is not a significant finding
- $p < .05$, there is less than a 5% chance that there is a significant finding
- $p < .01$, there is less than a 1% chance that there is a significant finding
- $p < .001$, there is less than a .01% chance that there is a significant finding

Figure 7.1 Alpha significance levels.

factors, including sampling error, researcher bias, alternative explanations, etc. Let's look at an example, not from social sciences, but from the pharmaceutical industry. If someone has a headache, and they decide to take a pill for their headache, the pill has gone through rigorous clinical trials. If Company X, who made that pill, set a confidence interval of 95% or an alpha significance level of $p < .05$, this means they are 95% certain the pill works and has no adverse effects in 95 out of 100 people. However, what about in a population of 100,000 people? The company, setting an interval of ($p < .05$), is stating that they are ok with potentially 5000 in every 100,000 people having adverse effects. It is HIGHLY doubtful that any regulatory body would approve this headache pill. What most companies do is raise the confidence interval threshold to something like 99.999% or ($p < .0001$). This means that 1 in every 100,000 could potentially have an adverse effect, which might still be high, but much better than ($p < .05$). Figure 7.1 shows four different alpha significance levels and what each means for determining if a statistical result is significant or not.

Summary

This chapter introduced you to the concept that renders making inferences with statistics possible: hypothesis testing. First, we discussed the importance of the use of a hypothesis in ensuring the falsifiability in scientific inquiries. We then delved into the rationale for NHST and delineated the differences between one-tailed versus two-tailed hypotheses. We have also used examples to illustrate the important elements in writing research hypotheses. Finally, the chapter explained the nature of Type I and II errors, and how researchers live and battle with them in everyday research life. And you now know what TTFN stands for too!

References

Kelly, S., & Gaytan, J. (2020). The effect of instructors' immediate behaviors and clarity on student writing apprehension. *Business and Professional Communication Quarterly*, *83*(1), 96–109. https://doi.org/10.1177/2329490619868822

Trump, D. J. (2017, January 27). *Executive order 13769: Protecting the nation from foreign terrorist entry into the United States*. Federal Register, 82 FR 8977.

Chapter 8

Correlation

What Will I Learn in This Chapter?

Under that white plaststeel armor, Stormtroopers, more than anyone else, could use a cold drink on a hot summer day. It seems that our craving for a cold drink intensifies on hotter days. In other words, as the temperature rises, so does our love for cold drinks. This relationship between temperature and our desire for cold drinks is highly intuitive to us and pervasive in everyday life. You might be tempted to claim: "That relationship is a positive correlation." Well, you are not wrong. However, correlation might be a slightly different concept when used in statistics, and it takes a lot more analysis before we can declare a significant correlation relationship between two variables. To this end, this chapter will cover:

- Covariance
- Correlation coefficient

DOI: 10.4324/9781003500117-8

- One-tailed versus two-tailed correlations
- How to run correlation in Excel and SPSS

Covariance

Imagine you are running a lemonade stand as a side business in the summer. It is important to figure out how many cups of lemonade you can sell because preparing too much means you must throw away or drink all unsold lemonade yourself and preparing too little will prohibit you from maximizing the profit. You have a hunch that people are more likely to enjoy a cup of cold, sweet lemonade on hotter days. Over the past week, you kept track of the temperature and the number of cups you sold each day. Below in Table 8.1 is what you found:

Just from seeing the numbers and without getting into any calculations, we can readily get a sense that the lemonade sales on hotter days are better. If we plot this data on a graph, with temperature on the x-axis and the number of cups sold on the y-axis, we will notice that as the temperature increases, so do our lemonade sales, as shown in Figure 8.1.

Table 8.1 Data of Lemonade Sales in a Summer Week

Day of the Week	*Temperature*	*Lemonade Sales*
Monday	25°C	20 cups
Tuesday	29°C	30 cups
Wednesday	33°C	28 cups
Thursday	32°C	50 cups
Friday	36°C	85 cups
Saturday	30°C	50 cups
Sunday	27°C	18 cups

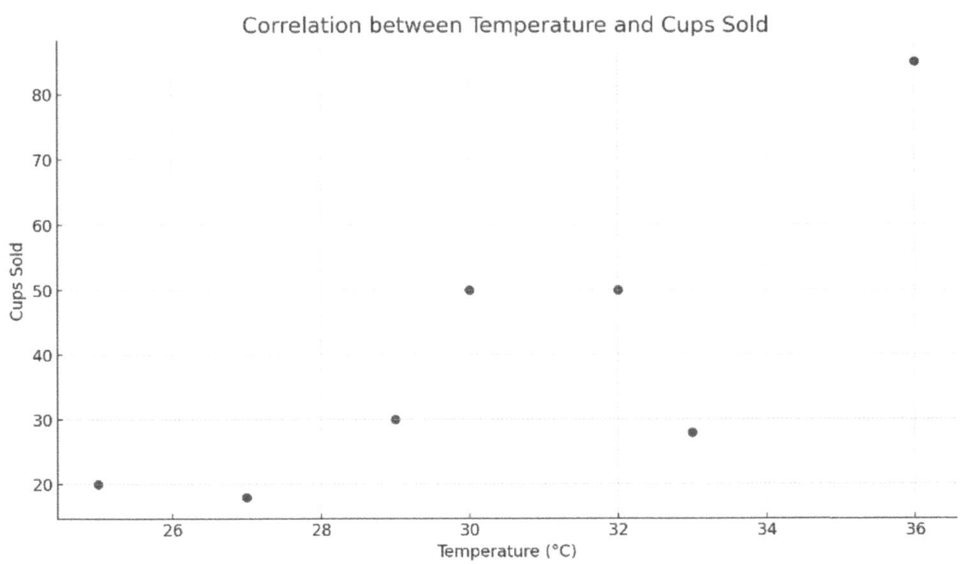

Figure 8.1 Scatter plot of temperature versus lemonade sales.

Table 8.2 Calculation of the covariance of temperate and lemonade sales

x_i	y_i	$x_i - \bar{x}$	$y_i - \bar{y}$	$(x_i - \bar{x})(y_i - \bar{y})$
25	20	−5.29	−20.14	106.47
29	30	−1.29	−10.14	13.04
33	28	2.71	−12.14	−32.96
32	50	1.71	9.86	16.90
36	85	5.71	44.86	256.33
30	50	−.29	9.86	−2.82
27	18	−3.29	−22.14	72.76
				$\Sigma = 429.71$

In this case, we have observed that the temperature and lemonade sales **covary** and we can calculate the **covariance** that measures the degree to which these two variables change together. Below is the formula for calculating the covariance between x and y:

$$Cov(x, y) = \frac{\sum(x_i - \bar{x})(y_i - \bar{y})}{n - 1}$$

x_i : values of the x-variable in a sample
$\bar{x}$: mean of the values of the x-variable
y_i : values of the y-variable in a sample
$\bar{y}$: mean of the values of the y-variable

In our example, x_i is the daily temperature, whereas y_i refers to the lemonade sales in cups. First, we need to calculate the means of both variables, and we will find that $\bar{x}$ = 30.29 and $\bar{y}$ = 40.14. You can then follow the steps outlined in Table 8.2 to calculate the covariance.

$$Cov(x, y) = \frac{429.71}{7 - 1} = 71.62$$

Here, a positive covariance value indicates that as temperature deviates from the mean, the lemonade sales would deviate in the same direction. While the covariance value provides valuable information on whether the two variables change in the same direction, the number 71.62 may appear quite arbitrary. Is it a larger or smaller number? If we sold thousands of lemonades in the previous example, we'd likely get a large covariance value. What does that mean? Additionally, how strongly are the two variables related anyway? If you are pondering these questions, you are on the right track to identifying the two main weaknesses of using covariance as a measure of relationship: the covariance value is (1) too dependent on the scales of measurement used and (2) does not provide information on the strength of the relationship. The good news is that we can overcome these weaknesses by using a standardized measure: **correlation**.

Test your understanding 8.1: What are the weaknesses of using covariance as a measure of relationship?

Correlation Coefficient

A correlation coefficient is the standardized measure of the relationship between two variables. We might frequently hear and use the word correlation in everyday life. In statistics, **linear correlation** is used to denote the relationship between two continuous variables and assumes that changes in one variable are associated with proportional changes in the other. A **positive relationship** means that both variables move in the same direction. It is still a positive relationship when one variable decreases, the other one also decreases. On the contrary, the variables are **negatively correlated** when they move in opposite directions. For example, the relationship between income and anxiety is typically a negative one, as we tend to feel more anxious when our income decreases. **Pearson correlation coefficient** (*r*) is the most common way of measuring a linear correlation between two variables, and below is the formula for calculating *r*:

$$r = \frac{\sum(x_i - \bar{x})(y_i - \bar{y})}{\sqrt{\sum(x_i - \bar{x})^2 \sum((y_i - \bar{y})^2}}$$

r : correlation coefficient
x_i : values of the x-variable in a sample
$\bar{x}$: mean of the values of the x-variable
y_i : values of the y-variable in a sample
$\bar{y}$: mean of the values of the y-variable

When comparing this to the calculation of the covariance in Table 8.2, we will notice the first steps are the same, but with a few additional calculations, as we need to know the values of $(x_i - \bar{x})^2$ and $(y_i - \bar{y})^2$. You can follow Table 8.3 if you want to get a taste of calculating the Pearson correlation coefficient by yourself.

$$r = \frac{429.71}{\sqrt{83.43 * 3352.86}} = 0.81$$

First, a positive *r* value indicates there is a positive correlation between temperature and lemonade sales. But how strongly are these two variables related? We can see the general trend of higher temperatures associated with higher sales, but with some exceptions. For instance, Sunday sales are lower when compared to Monday, despite Sunday being slightly hotter. Here, a positive relationship between temperature and lemonade sales does not guarantee that these two variables move consistently together. This might allude to the importance of understanding the strength of a

Table 8.3 Calculating the Pearson Correlation Coefficient between Temperature and Lemonade sales

x_i	y_i	$x_i - \bar{x}$	$y_i - \bar{y}$	$(x_i - \bar{x})(y_i - \bar{y})$	$(x_i - \bar{x})^2$	$(y_i - \bar{y})^2$
25	20	−5.29	−20.14	106.47	27.94	405.73
29	30	−1.29	−10.14	13.04	1.65	102.88
33	28	2.71	−12.14	−32.96	7.37	147.45
32	50	1.71	9.86	16.90	2.94	97.16
36	85	5.71	44.86	256.33	32.65	2012.16
30	50	−.29	9.86	−2.82	.08	97.16
27	18	−3.29	−22.14	72.76	10.80	490.31
				$\sum$ =429.71	$\sum$ =83.43	$\sum$ =3352.86

Table 8.4 Interpretation of Pearson Correlation Coefficient *r*

r value	*Interpretation of r value*
$r = 1$, or $r = -1$	Perfect correlation
$0.7 \le r < 1$, or $-1 < r \le -0.7$	Strong correlation
$0.5 \le r < 0.7$, or $-0.7 < r \le -0.5$	Moderate correlation
$0.3 \le r < 0.5$, or $-0.5 < r \le -0.3$	Weak correlation
$-0.3 < r < 0.3$	Little to no correlation

correlation (**effect size**) that shows how closely related these two variables are. Luckily, the standardized covariance value, *r*, lies between −1 and +1 and provides important information on the effect size. In Table 8.4, you will find a breakdown of interpretations of different *r* values.

The strength of the correlation provides important insights into how consistently the two variables move in a predictable way. Take active listening, for example: When we receive active listening responses, we tend to feel more satisfied with the conversation and find the communicator more effective and attractive (Weger Jr. et al., 2014). A strong correlation between active listening and perceived attractiveness suggests that those who exercise active listening are more likely to be perceived as attractive in social settings. On the other hand, the relationship between the use of emojis and perceived likability is often considered a weak correlation because the context matters. In other words, while we know correctly that using emojis in conversations can help enhance our social image, the effect is likely limited. When we get to the chapter on regression analysis, the strength of the correlation will be made clearer as we will learn how it is closely tied to the concept of **explained variance**.

In our lemonade stand example, our *r* value is 0.81, suggesting a strong, positive correlation between the temperature and lemonade sales. You can formally report this correlation in a study as follows: A positive relationship was found between temperature and lemonade sales, with a correlation coefficient of $r = 0.81$. When reporting a correlation, we need to mention the variables, the direction of the relationship, the *r* value, and the *p*-value. For instance, in a study exploring workplace rumination and employee dissent, Zeng et al. (2022) reported that negative rumination was positively associated with articulated dissent [$r(395) = .28$, $p < .001$]. In Shimotsu-Dariol et al. (2012)'s study, the authors found that students' academic competitiveness was related positively to their out-of-class communication with instructors ($r = .18$, $p < .01$) and their in-class participation ($r = .17$, $p < .01$). It is common in social sciences that scholars would perform correlations among all studied variables and present them in a correlation table as basic information for readers. You can find more examples in Shimotsu-Dariol et al. (2012) and Zeng et al. (2022) on how to report a correlation and draw a correlation matrix. In fact, you should be able to find a correlation matrix in most quantitative studies.

Typically, both variables need to be normally distributed continuous data for a Pearson's correlation to be correctly performed. One exception is that one of the variables can be a categorical variable but with only two categories (i.e., dichotomous variable, usually a dummy variable). When running a Pearson's correlation in this case, we are essentially performing a t-test, which will be covered in more depth in Chapter 10. You will also find more information about tests involving a dummy variable in the latter chapters on regression analyses.

<u>Test your understanding 8.2:</u> The *r* value for the correlation between TV-watching hours and student grades is −0.43 with $p < .05$. How would you interpret this result?

One-Tailed or Two-Tailed Correlations

When performing a correlation test, we are faced with the decision between a one-tailed and two-tailed correlation. That decision is dependent on whether you have a specific directional hypothesis about the relationship between the two variables or if you're simply interested in determining if a relationship exists without assuming directionality.

One-Tailed Correlation

A one-tailed test is used when you hypothesize a specific direction for the relationship. In this case, if you expect that higher temperatures will lead to an increase in lemonade sales, you could set up a one-tailed test. Keep in mind that when running a one-tailed correlation, all the significance level is allocated to one end of the distribution because you're only looking for a relationship in one direction (positive or negative).

Two-Tailed Correlation

A two-tailed test is used when you do not assume a specific direction of the relationship, meaning you're open to both possibilities: either an increase or a decrease in sales as temperature changes. This approach is generally more conservative and is suitable when the relationship's direction is uncertain or when you just want to see if a correlation exists in either direction. Compared to one-tailed correlation, it requires more extreme values to achieve significance because the significance level is split on both ends. You can think of it this way: we generally require $p < 0.5$ to declare statistical significance. In a two-tailed test, we would need to have $p < 0.25$, as we are testing significance from both ends. In other words, two-tailed correlation is a more conservative approach that is less likely to make Type I errors.

Test your understanding 8.3: You are confident that one tends to have better communication skills after taking more communication classes. Would you run a one-tailed or two-tailed correlation here?

Correlation and Causation

Correlation and causation are two closely related but principally different concepts. Unlike correlation, causation implies that one variable directly affects or changes another. For instance, the number of study hours likely has a positive, direct impact on your exam scores. Causation is typically harder to establish, as we need to rule out other factors. This is also why controlled experiments are very common in the process of drug development when researchers want to establish that the drug, instead of some random factors, has a direct influence on the illness. With a significant correlation result, we can merely conclude that the variables co-occur in a certain way. For a good laugh, you can visit the site www.tylervigen.com/spurious-correlations for some spurious correlation examples, such as yogurt consumption being strongly correlated with UFO sightings. It is surely silly for one to mistake a correlation for a causation here.

Curvilinear Relationships

While we are predominantly focusing on linear relationships in this chapter, it is highly common that two variables are related, and their relationship does not follow a straight line. Instead, the relationship

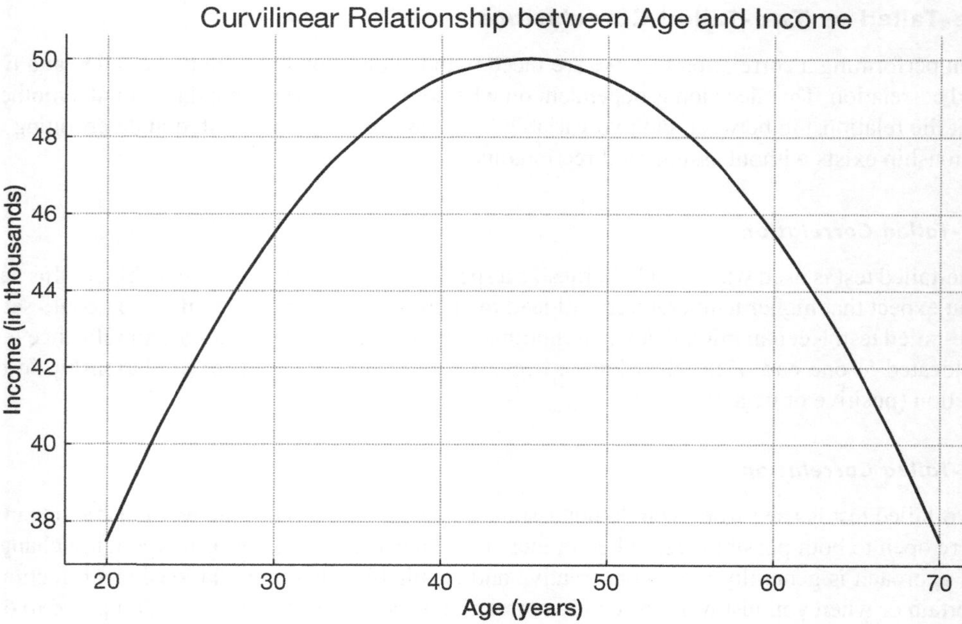

Figure 8.2 A curvilinear relationship between age and income.

forms a curve when plotted on a graph. For example, age and income are two closely related variables as income often increases when young people gain experience and reach career peaks, but it can level off or decline as people approach retirement (see Figure 8.2). Similarly, stress level might have a curvilinear relationship with performance, where some stress can improve performance, but too much may destroy it. It is important to keep in mind that curvilinear relationships are highly common in real-world scenarios where the effect of a variable may increase, peak, and then decline. Thus, a non-significant linear correlation coefficient is not sufficient for us to conclude that these two variables are not "related" because there is still a chance that they might be significantly related in a curvilinear way.

Spearman Correlations

So far, everything covered in this chapter refers specifically to Pearson correlations. While Pearson's correlation is commonly used for measuring linear relationships between two continuous variables, it assumes that the data is normally distributed and have a linear relationship. As you've seen in prior examples from this book, not all relationships in real-world data meet these assumptions.

Spearman's rank-order correlation is a non-parametric alternative that assesses the strength and direction of a monotonic relationship between two variables. A **monotonic relationship** is one in which both variables increase or decrease together, but not necessarily at a constant rate. Spearman correlations are useful when dealing with ordinal data or nonlinear relationships. For example, if you wanted to correlate class rank (First Year, Sophomore, Junior, Senior) with public speaking anxiety, Spearman's correlation would be more appropriate than Pearson's because class rank is ordinal. The value of Spearman's rho (ρ), like Pearson's r, ranges from -1 to $+1$, with values closer to ± 1 indicating stronger relationships and values closer to 0 indicating weaker relationships.

Running Pearson's Correlation in Excel

There are two methods for running a correlation in Excel. For the first method:

1. Prepare the data (e.g., x_i and y_i) in two columns
2. Select an empty cell and use the CORREL formula as shown in Figure 8.3, where you type =CORREL() with the ranges of the two variables listed in parentheses, separated by a comma
3. Press ENTER, and you will get the correlation coefficient between the two sets of data

For the second method:

1. Make sure you have prepared the data the same way as in method 1
2. Open ToolPak (install it if you haven't already) and find Correlation, as shown in Figure 8.4

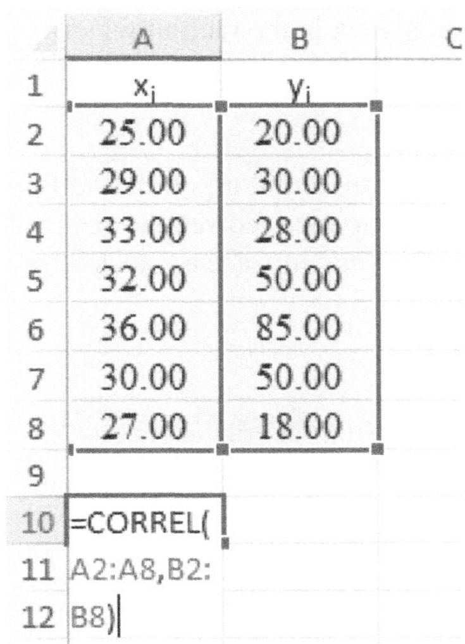

Figure 8.3 Correlation in Excel using method 1.

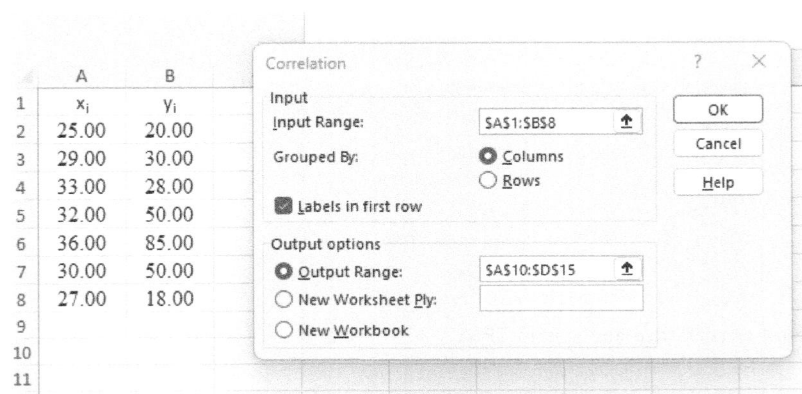

Figure 8.4 Correlation in Excel using ToolPak.

3. For Input Range, select the entire array of data, which in this case is A1 to B8
4. For Output Range, identify a blank cell in the spreadsheet
5. Click OK

Running Pearson's Correlation in SPSS

Once data is organized into two columns in SPSS, navigate to:

1. Analyze > Correlate > Bivariate, as shown in Figure 8.5

From there, move the variables of interest into the Variables column, as shown in Figure 8.6. Pearson and Two-tailed are being selected as the default. If you have a directional hypothesis in mind, you can change it to One-tailed. A two-tailed correlation is run here for simplicity reasons. Click *OK*.

This will produce a correlation matrix in the Output window as shown in Table 8.5.

Summary

In this chapter, we have discussed correlation, a popular and powerful statistical tool for understanding patterns in data. Unlike covariance, correlation isn't affected by the scales of the variables and ranges between −1 and 1, making it easy to interpret across different types of data.

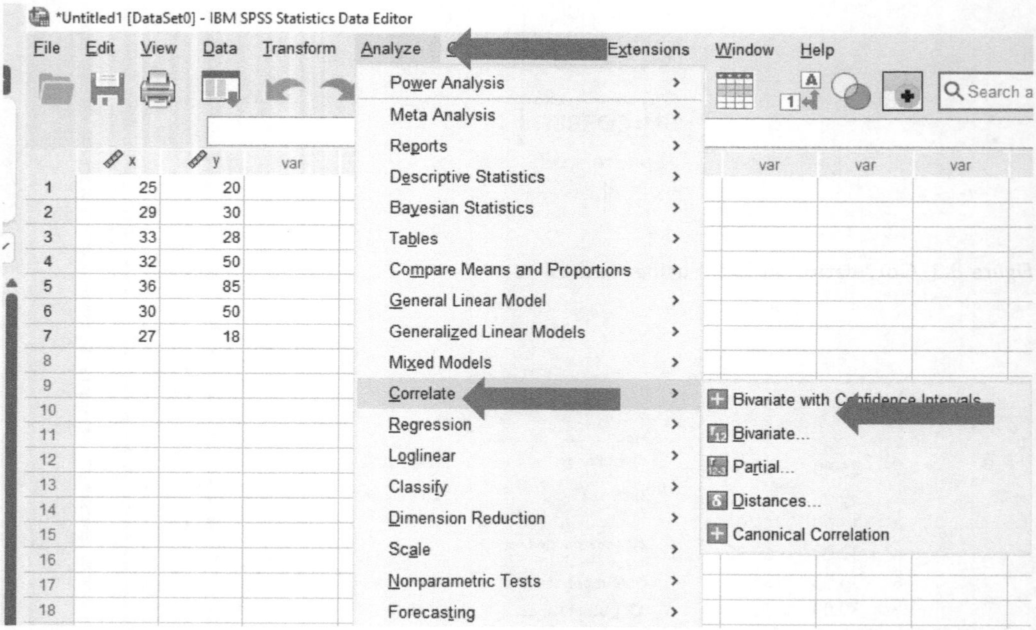

Figure 8.5 Find correlation analysis in SPSS.

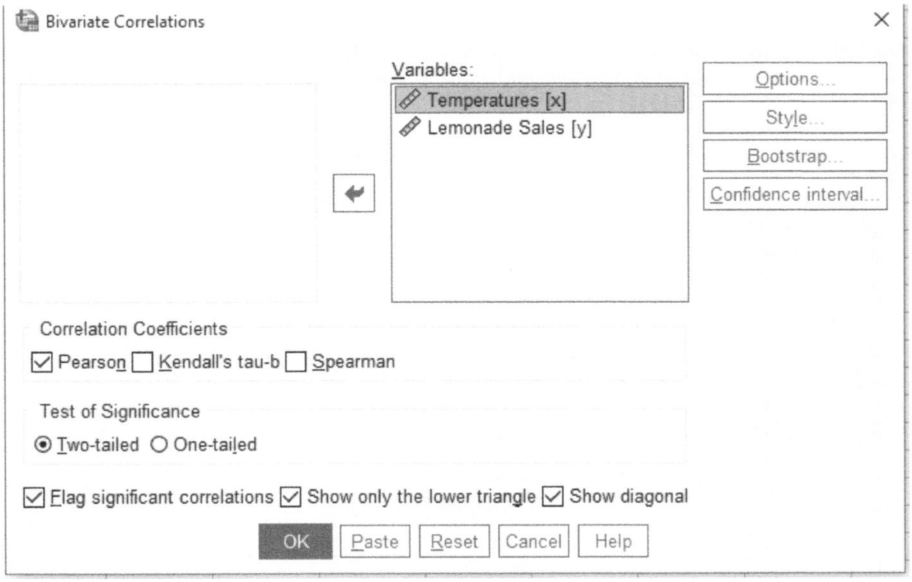

Figure 8.6 Correlation analysis specification in SPSS.

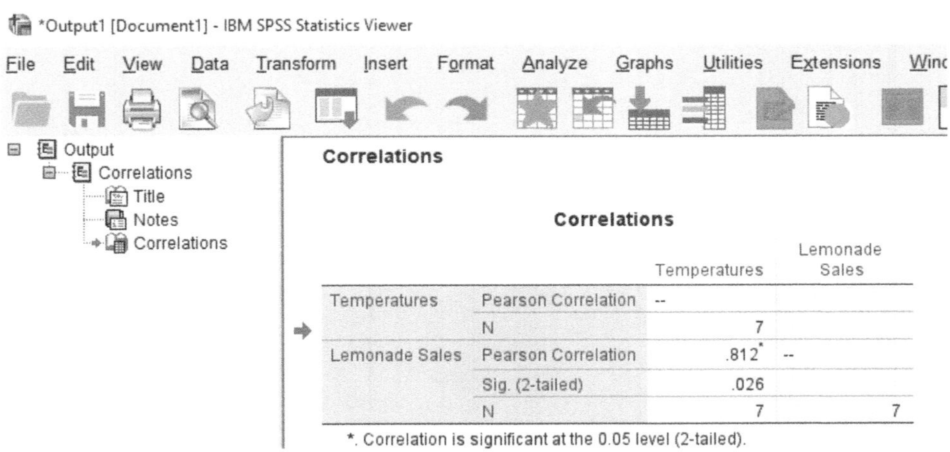

Correlations

		Temperatures	Lemonade Sales
Temperatures	Pearson Correlation	--	
	N	7	
Lemonade Sales	Pearson Correlation	.812*	--
	Sig. (2-tailed)	.026	
	N	7	7

*. Correlation is significant at the 0.05 level (2-tailed).

Table 8.5 Correlation Output in SPSS

It is important to avoid making causal claims when interpreting correlation results. Now, you understand the differences between running a one-tailed and two-tailed correlation. You've also learned different ways of performing a correlation analysis as well as how to correctly interpret the significance and strength of a correlation.

References

Shimotsu-Dariol, S., Mansson, D. H., & Myers, S. A. (2012). Students' academic competitiveness and their involvement in the learning process. *Communication Research Reports*, *29*(4), 310–319. https://doi.org/10.1080/08824096.2012.723643

Weger Jr., H., Castle Bell, G., Minei, E. M., & Robinson, M. C. (2014). The relative effectiveness of active listening in initial interactions. *International Journal of Listening*, *28*(1), 13–31. https://doi.org/10.1080/10904018.2013.813234

Zeng, C., Rahmani, D., & Okafor, B. (2022). An investigation of the mediating role of organizational trust in the relationship between work rumination and organizational dissent. *Communication Studies*, *73*(4), 441–457. https://doi.org/10.1080/10510974.2022.2089190

Chapter 9

Simple Linear Regression

What Will I Learn in This Chapter?

Dorian is an avid Lego builder. He has a design idea for a new Lego city of his own creation, but he doesn't know how many Legos he will need to manifest his vision. He starts building some of the walls he sees in his mind that he knows he has enough pieces of the right size, shape, and color to construct. Then he takes stock of the number of bricks he used for each wall:

- Wall 1: 2 bricks high, 11 bricks used
- Wall 2: 4 bricks high, 19 bricks used
- Wall 3: 6 bricks high, 31 bricks used
- Wall 4: 8 bricks high, 40 bricks used

DOI: 10.4324/9781003500117-9

He notices a pattern: there are almost five times as many bricks in each wall as there are bricks in the wall's height. This means he can estimate how many bricks he will need for each wall using this equation:

$$Y = 5X$$

Here Y is the total number of bricks needed, and X is how many bricks high the wall should be.

Is this an exact calculation? No! Using this equation to calculate the number of bricks used per wall would underestimate Walls 1 and 3, but overestimate Wall 2. The only wall that would be perfectly explained in this equation is Wall 4. It is, however, the equation that gets closest to estimating all the wall heights. Therefore, it is the best choice for Dorian to use to estimate his building material needs, even though he may be off a brick or two in his estimates. Using this equation to help himself, Dorian is using simple linear regression; he has found the line that is the best fit for estimating how many bricks he will need per wall, depending on how high he wishes the wall to be. In this chapter, you will:

- Understand what simple linear regression is
- Review the strengths and weaknesses of simple linear regression
- Understand when correlation versus simple linear regression is the appropriate analysis
- Learn how to run a simple linear regression in Excel and SPSS

What Is Simple Linear Regression?

Simple linear regression is a statistical method that shows the relationship between two variables. This may feel extremely familiar to what you read in Chapter 8 about correlation. In fact, the two methods are very similar. Recall that in Chapter 8, you read that correlation is a statistical method for finding the relationship between two variables when you have no evidence that one is causing a change in the other, only that they trend together. Simple linear regression looks for the same information about the trending of variables but holds an underlying assumption that one variable (i.e., independent variable) causes a change in another (i.e., dependent variable).

Simple linear regression is calculated using the following equation model:

$$Y = \beta X + a + \varepsilon$$

where Y is the dependent variable
X is the independent variable
β is the slope of the line,
a is the y-intercept
ε is the error in the line

This simple linear regression equation may look very familiar to you, not just from the equation in the introduction but also from your algebra class. Do you recall algebra using the following formula?

$$Y = mX + b$$

In algebra, you were taught that Y is the dependent variable, X is the independent variable, m is the slope of the line, and *b* is the y-intercept. The idea is the same, and the fact that one uses *β* and *a*

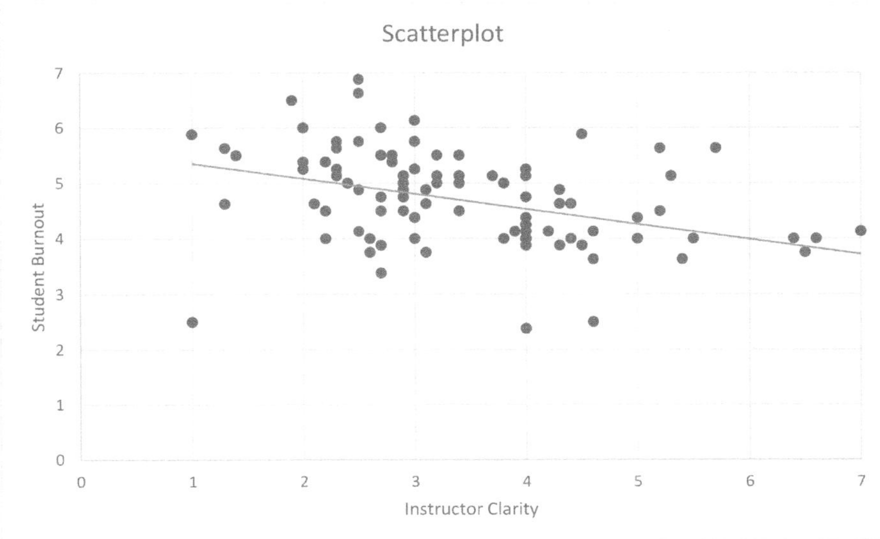

Figure 9.1 Scatterplot.

versus m and b is simply so you know whether you're working on an algebraic problem with exact calculations or a statistical problem in which you are calculating the closest possible estimate.

Recall also in Chapter 4 that you learned about scatterplots. In that example, you saw the intersection of burnout and clarity measures using 20 data points rounded to the nearest whole number. Here is that scatterplot representing student burnout and instructor clarity again in Figure 9.1 with the entire 100-row dataset of the raw, unrounded data. In addition to having the full dataset, Figure 9.1 also differs from the scatterplot you saw in Chapter 4, in that it has a line running through it. This is a **trendline**, a line of best fit. The trendline is a visual representation of a regression equation. Regression is calculated to find the trendline that passes through a scatterplot that is the closest line possible to all the points.

Assumptions

Notice that in Figure 9.1, very few points fall on a regression line (i.e., trendline), but it is the line of best fit that best predicts the position of burnout as a product of clarity. Here, you see the difference between regression and an algebraic line: in algebra, all points would fall on the line, but in regression, very few points will, yet the line will capture the general slope of points. In regression, the closer the points are to the line, the better the fit of the line, and therefore the more accurate the regression equation. Herein is the first assumption of regression: there is a linear relationship between variables. The linear relationship does not imply a perfect line, but rather that the movement of datapoints can be regressed to a similar slope. This assumption can be checked by looking at the scatterplot to see whether the data tends to trend in a linear fashion.

Regression also assumes observations are independent. This simply means you cannot use the same measure to represent both independent and dependent variables. This may seem like common sense now, but it was not always so in the early days of social science measurement, so it's worth documenting as an assumption today. We can imagine how this assumption may be violated through a thought exercise with the variable source credibility. Source credibility has three dimensions: caring, competence, and trustworthiness (McCroskey & Teven, 1999). The three dimensions can be examined separately or used together as a composite assessment of credibility. Imagine you were to use a simple linear regression to see how the subdimension of competence (independent variable) predicts the composite measure of credibility (dependent variable). This would violate the assumption of independence because credibility is measured in part by competence, and as such is not a totally independent measure. This assumption of independence is managed entirely in the design of the data collection.

The third assumption is that of **homoscedasticity**. This is an unreasonably difficult word to pronounce that simply means that for the regression line to be a decent predictor, there should be roughly as many points above it as below it in a scatterplot. Homoscedasticity is checked by glancing at the scatterplot. Homoscedasticity is affected by the presence of extreme outliers that may skew the line of fit away from the center of the data. When a dataset has homoscedasticity issues, sometimes outlier data needs to be excluded from the analysis for the regression line to be a decent predictor.

The fourth assumption is normality of residuals. This means if you created a histogram of this distance from each datapoint to the regression line, then the histogram would have a normal distribution. So again, this assumption can be tested visually.

Finally, simple linear regression assumes no autocorrelation. This is a concern primarily for data collected at different points in time. It means that simple linear regression analyses assume that there is nothing happening across time that influences the data (e.g., if one of your variables is temperature, you cannot collect temperature data back-to-back because the temperature one day influences the temperature the next day). As such, autocorrelation is similar to the assumption of independence of variables, except it pertains to allowing enough time to pass between observations that variables are showing independent observations rather than observations still representing prior data. You'll learn more about testing for autocorrelation in the next chapter.

Test Your Understanding 9.1 – What must be true about your data for simple linear regression to be possible as an analysis?

Interpretations

Now that you have looked at the trendline representing the regression equation in the burnout and clarity dataset, it's time to see the regression equation itself. The equation for this line is:

$$Y = -.55X + 6.05$$

(You'll see how to get that equation later.) Right now, it's time to focus on what the equation means. This tells us there is a negative relationship between the variables because the coefficient ($\beta = -.55$) that determines the slope of the regression line is negative. The line uses an **unstandardized coefficient** for slope. This means that for every increase of 1 in the independent variable (instructor clarity), the dependent variable (student burnout) will go down by.55 units in a visual depiction. It should also be noted that the equation for the line may be written as

$$Y = -.55X + 6.05 + \varepsilon$$

Here, the equation acknowledges that while the line is drawn without error, it has error and is not a perfect fit to the data.

Okay, so now it's time to look at where the equation came from. See the regression output in Table 9.1 from SPSS. The y-intercept comes from the Constant, while the slope of the line (coefficient) comes from the Unstandardized Beta. Both are found under the Coefficients subheading. Note also that an estimate of the error is given under this subheading.

Regression coefficients are expected, like correlation coefficients, to have a value between −1 and 1. While −.55 falls within this range, that is not always the case for the equation of trendlines. This is why the beta for graphing a trendline is called the "unstandardized" coefficient. To interpret the magnitude of a relationship between two variables in simple linear regression, the unstandardized coefficient is adjusted from the full natural scale of values to a value between −1 and 1.

In Table 9.1, the standardized beta is −.39, and that is the value you interpret to understand the magnitude of the relationship between variables. That means, like when you talked about relationships in Chapter 8, the relationship is negative: as one variable goes up, the other goes down and vice versa. Also, as you learned in Chapter 8, the closer the value is to zero, the weaker the relationship, while the closer to −1 or 1, the stronger the relationship. Here, there is a weak negative relationship between clarity and burnout. The −.39 (or −.389) came from the **standardized coefficient** beta, which has been transformed to scale, so it ranges between −1 and 1.

Also familiar to you from Chapter 8 is the significance value. Using the Sig. value associated with the standardized beta coefficient ($p < .001$), you can see the regression coefficient is statistically significant because the significance level (p-value) is less than .05. Interpret this to mean the moderate negative relationship you see between variables likely did not happen by chance.

Table 9.1 Regression Output

Model Summary

Model	*R*	*R Square*	*Adjusted R Square*	*Std. Error of the Estimate*
1	.389[a]	0.151	0.142	1.1297

a. Predictors: (Constant), Clarity

Coefficients[a]

Model		*Unstandardized Coefficients*		*Standardized Coefficients*	*t*	*Sig.*
		B	*Std. Error*	*Beta*		
1	(Constant)	6.059	0.632		9.595	0.000
	Clarity	−0.554	0.133	−0.389	−4.177	0.000

a. Dependent Variable: Burnout

ANOVA[a]

Model		*Sum of Squares*	*df*	*Mean Square*	*F*	*Sig.*
1	Regression	22.265	1	22.265	17.45	<.001[b]
	Residual	125.065	98	1.276		
	Total	147.330	99			

a. Dependent Variable: Burnout
b. Predictors: (Constant), Clarity

The unstandardized and standardized beta coefficients, sometimes called **beta weights**, are often confusing to first-time statistics students because they are both representative of the same relationship, but you use one to graph the trendline and another to interpret the relationship's magnitude. The most simplistic explanation for thinking about how the two relate is to consider that when regression equations are calculated, the computer calculates all the distances between data points and all possible lines that could pass between those points until it finds the line of best fit. Then it provides a line using the exact values of estimated distances between points (i.e., the unstandardized coefficients). Quite often, though, these values are beyond the bounds of −1 and 1. The unstandardized values are smashed into standardized coefficient values so that all regressions can be interpreted with betas ranging from −1 to 1. That standardization allows us to compare the relationship between variables, no matter if the data range from 1 to 5 or 1 to 100. So, the purpose of using the standardized coefficient is to allow you to use the same procedure for assessing the magnitude of the relationship, no matter what range of scores your data may have.

A lot of what you've been told here is just to help you see the big picture behind what is happening when you tell Excel or SPSS to perform a simple linear regression. What you need to know when you run a simple linear regression is this:

1. What is the standardized beta?
2. What is the p-value of that standardized beta?

If you know both things, you know the magnitude and polarity of the relationship and know whether the relationship between your independent and dependent variables is statistically significant.

There is a second item to pay attention to in Table 9.1. Look at the Adjusted ***R*-Squared** (R^2adj) value under the Model Summary subheading. This adjusted *R*-squared (R^2adj) is the percentage of student burnout behavior that is explained by instructor clarity. So, we interpret this to mean that of all the things that affect how much burnout a student feels, 14% of it is explained by how clear their instructor is. This R^2adj is calculated by taking the *R* value and adjusting for error in the line of fit. That *R* value is the exact same as the *r* value you learned about in Chapter 9 on correlations. In fact, if you calculate the correlation between these two variables, $r = .39$, $p < .001$, which means *r* is the same as the standardized beta weight.

There is one last thing from the SPSS output shown in Table 9.1 that needs to be addressed: the ANOVA. A lot of this part of the table will not make much sense until Chapter 12, but it is worth taking a moment to consider the *F*-value. An *F*-test relative to a regression, tests whether the variance explained by the regression model (i.e., the line) is greater than the variance still unexplained. If more variance is explained than not, then the *F*-value will be statistically significant, as seen in Table 9.1. The *F*-value will always be statistically significant when the *p*-value for the independent variable is statistically significant in a simple linear regression. In terms of simple linear regression, the *F*-test is a different way of verifying that the independent variable is a predictor of the dependent variable. In the social sciences, the *F*-value is not reported with simple linear regressions results. However, it is useful for overall understanding to know why an ANOVA table is provided in the output of a simple linear regression.

Test Your Understanding 9.2 – Which values must you look at to interpret the results of simple linear regression? What do each of those values tell you?

Choosing Correlation versus Simple Linear Regression

In sum, simple linear regression does not give you much more information than correlation. The output will show you both standardized and unstandardized fit, whereas correlation is shown in the standardized output. Simple linear regression will also give you an R^2 value, whereas correlation will not. However, you can calculate the R^2 value in a correlation by simply squaring the correlation coefficient (r). So, the output is very similar. In fact, correlation is quite literally a two-variable regression without an estimate of error. Why then, if the two are so similar, would you choose one over the other?

The answer all boils down to having theoretical support for causation. If research is being guided by a theory or theoretical model, which indicates one of the variables causes change in the other, then the appropriate choice is simple linear regression. If there is no theoretical support to believe one variable causes change in the other, then the proper analysis is correlation.

The data examined in this chapter considers instructor clarity and student burnout. The Instructional Beliefs Model predicts that instructor behaviors (e.g., clarity) cause affective changes (e.g., burnout) in students within their classrooms (Weber et al., 2011). As such, there is a theoretical model that predicts instructor clarity causes a change in student burnout. This makes the appropriate analysis for these two variables simple linear regression, predicting that instructor clarity is the independent variable and student burnout is the dependent variable.

Consider what would happen if the data were used in a simple linear regression with the independent and dependent variables swapped. Table 9.2 shows the output of the regression with the X and Y variables confused. Note that the information used to plot the trendline has changed because the variables assigned to the x and y axes are different. However, the information you interpret [Adjusted R^2, Standardized Beta Coefficient, and Significance (p-value)] is all identical. As such, we interpret the same information (i.e., moderate negative relationship, statistically significant) even if the dependent and independent variables have been confused. This means the R value

Table 9.2 Regression Output Variables Reversed

Model Summary

Model	*R*	*R Square*	*Adjusted R Square*	*Std. Error of the Estimate*
1	.389[a]	0.151	0.142	0.79206

a. Predictors: (Constant), Burnout

Coefficients[a]

Model		*Unstandardized Coefficients*		*Standardized Coefficients*	*t*	*Sig.*
		B	*Std. Error*	*Beta*		
1	(Constant)	5.625	0.240		23.485	0.000
	Burnout	−0.273	0.065	−0.389	−4.177	0.000

a. Dependent Variable: Clarity

produced with the confused dependent and independent variable is also the same as the *r* value you would receive from a correlation.

So, why bother having an entire chapter on simple linear regression if it is so closely aligned to correlation that you will get the same answer no matter which analysis you use and no matter whether you get independent or dependent variables confused? Why not just teach one or the other when the only real difference is that for one, you should be particular about which is graphically depicted on the x-axis versus the y-axis, and the other does not care which is which? The answer is simply, scaffolding for learning. Your authors cannot recall a single social science paper that used simple linear regression to test a hypothesis because research is rarely so simple that one variable is expected to affect another and that is all a study examines. You are learning about simple linear regression because it is the link, the scaffolding, between correlation and your next chapter: multiple regression. Understanding this link between the two concepts makes the next chapter far easier to digest.

Test Your Understanding 9.3 – When would you choose correlation versus simple linear regression to analyze your data?

Simple Linear Regression in Excel

Running a simple linear regression in Excel takes place in the Data Analysis tool that you're familiar with. From that menu, click "Regression" as shown in Figure 9.2, then OK.

From there fill in the information indicated in Figure 9.3:

1. Fill in the "Input Y Range" blank with the cell range that holds the values of your dependent variable
2. Fill in the "Input X Range" blank with the cell range that holds the values of your independent variable
3. Ensure that your Confidence Interval is set to 95%
4. Indicate in "Output Range" the cell you wish for your first column and row of output to be printed in
5. Click any additional plots you wish to be produced

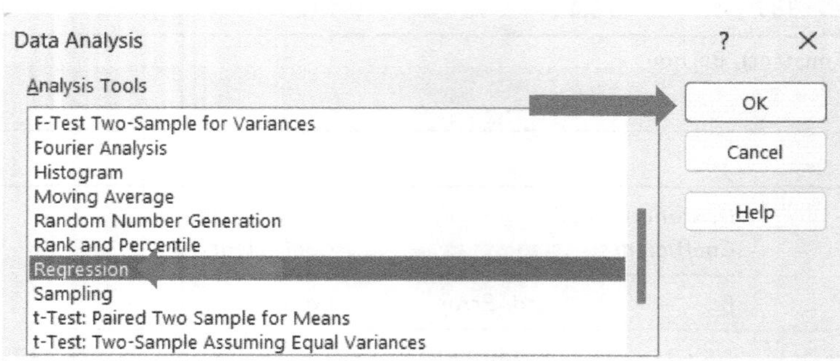

Figure 9.2 Data Analysis.

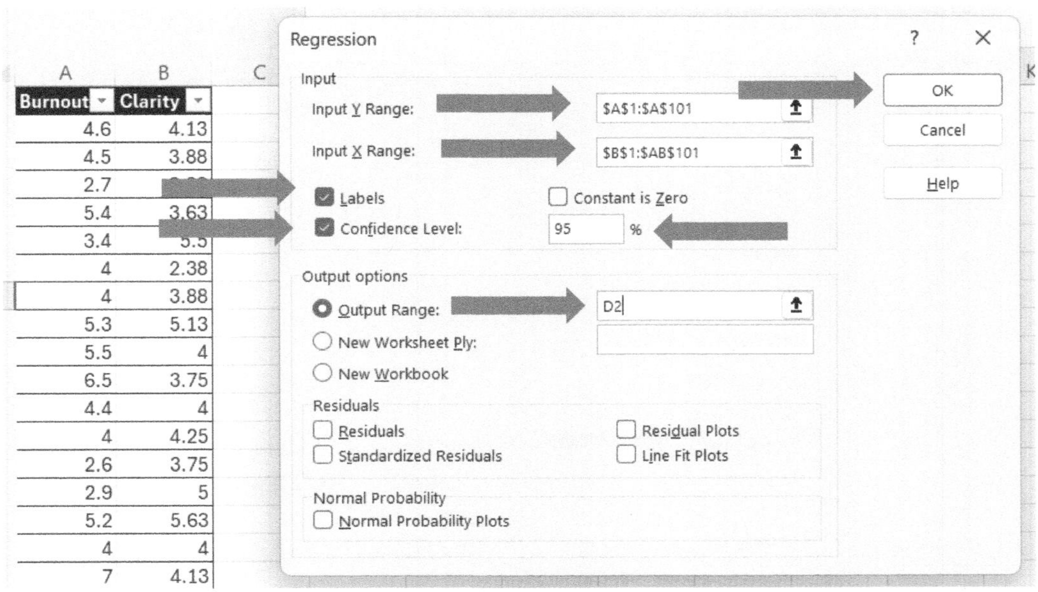

Figure 9.3 Input and Output.

Simple Linear Regression in SPSS

Computing a simple linear regression takes almost as few steps as Excel. First, as shown in Figure 9.4, open the regression menu by clicking:

1. Analyze
2. Regression
3. Linear . . .

These steps will open the window for the regression tools shown in Figure 9.5. Within this window:

1. Click on the independent variable and then the arrow next to Block 1
2. Click on the dependent variable and then the arrow next to Dependent
3. Click OK

In the example used in this chapter, the data fit well to a straight line. Data will not always do that. The name of this analysis is "linear" regression. Not all lines are straight though. You may have data that takes the shape of a quadratic, exponential, or logarithmic line (see Figure 9.6). In fact, you may have data that takes the shape of any sort of line that you remember from your math classes.

To find the shape of your data, build a scatterplot. Remember, Chapter 4 shows you the steps for building a scatterplot. When you look at a scatterplot, if your data does not seem to regress to a straight line, then do not force it to do so. SPSS will allow you to regress to any line type that your data seems to best fit.

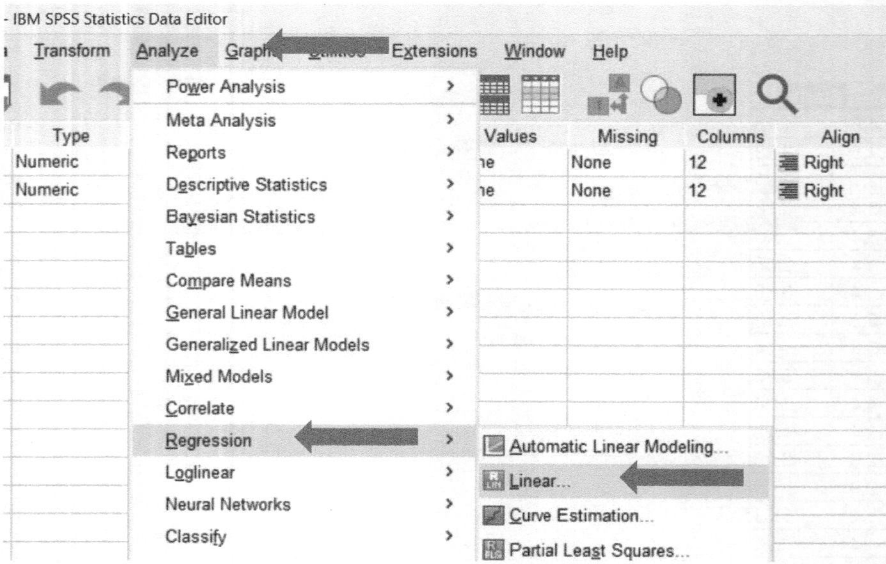

Figure 9.4 SPSS Regression menu.

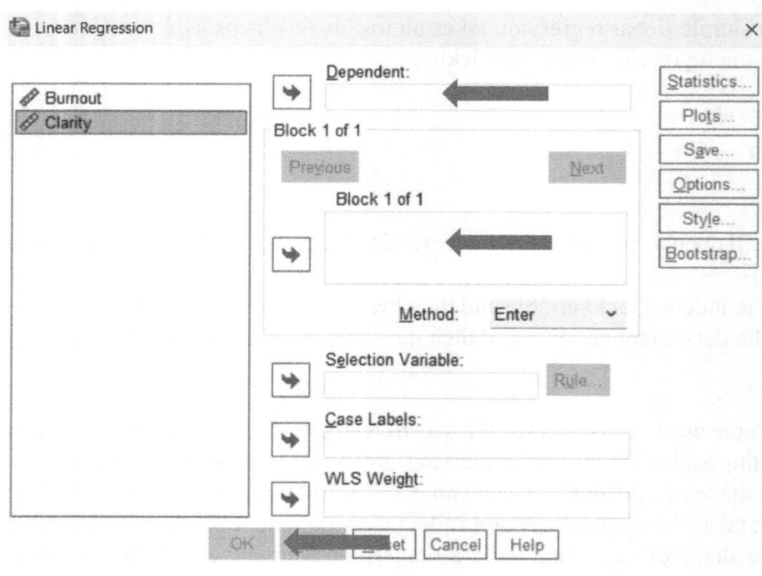

Figure 9.5 SPSS Regression window.

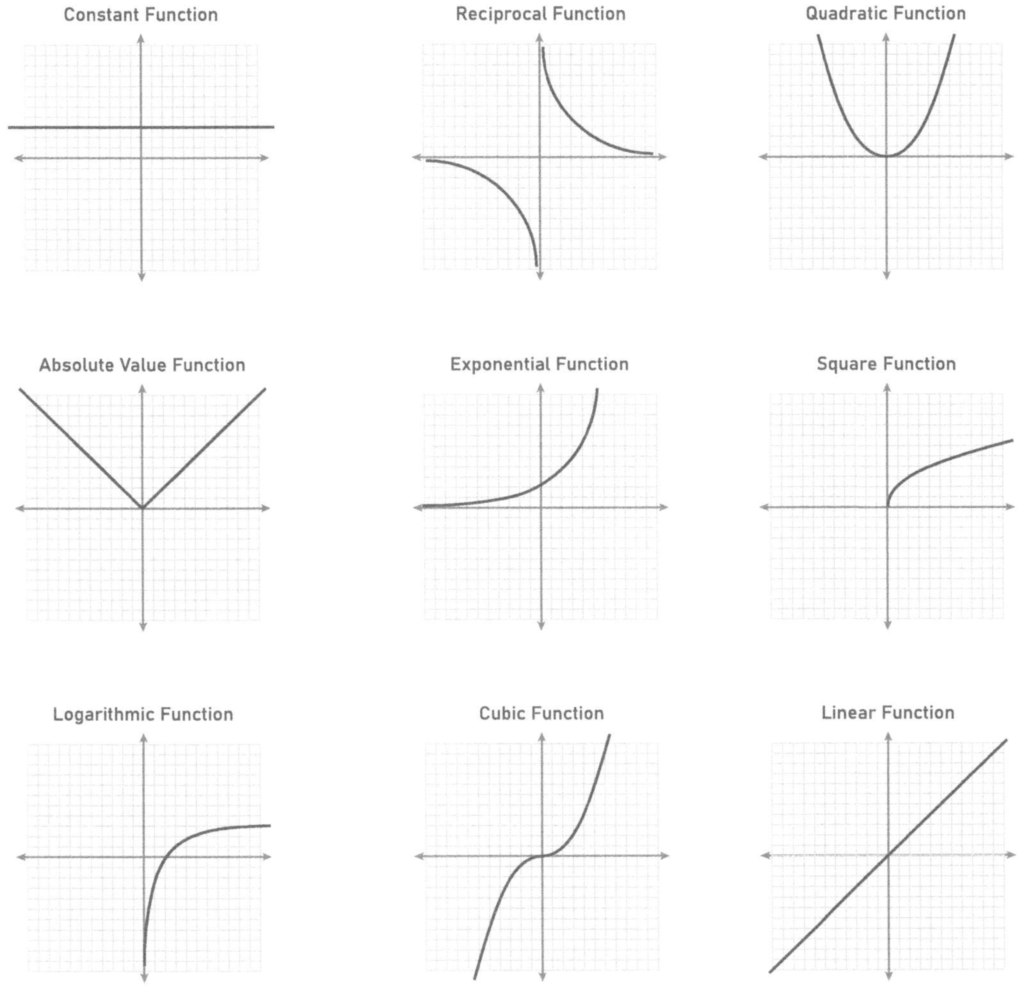

Figure 9.6 Line types.

If you need to build a regression line that is curved, then to do that in SPSS, you begin your analysis differently. To begin, go to (see Figure 9.7):

1. Analyze
2. Regression
3. Curve Estimation . . .

Once Curved Estimation. . . is selected, the box shown in Figure 9.8 will appear. From here:

4. Insert your independent variable in the independent variable box and dependent variable in the dependent variable box using the same steps as shown earlier, clicking on your variables and moving them into the appropriate boxes with the arrows

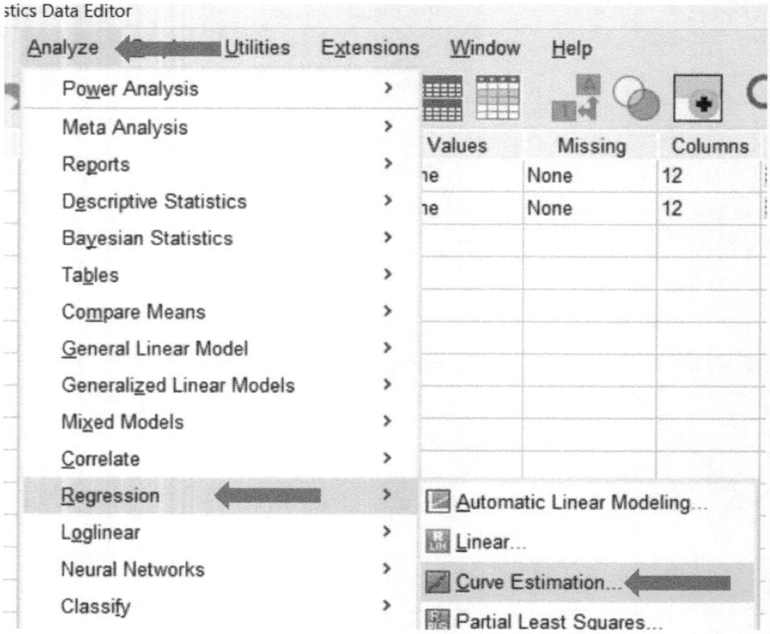

Figure 9.7 Curved regression line.

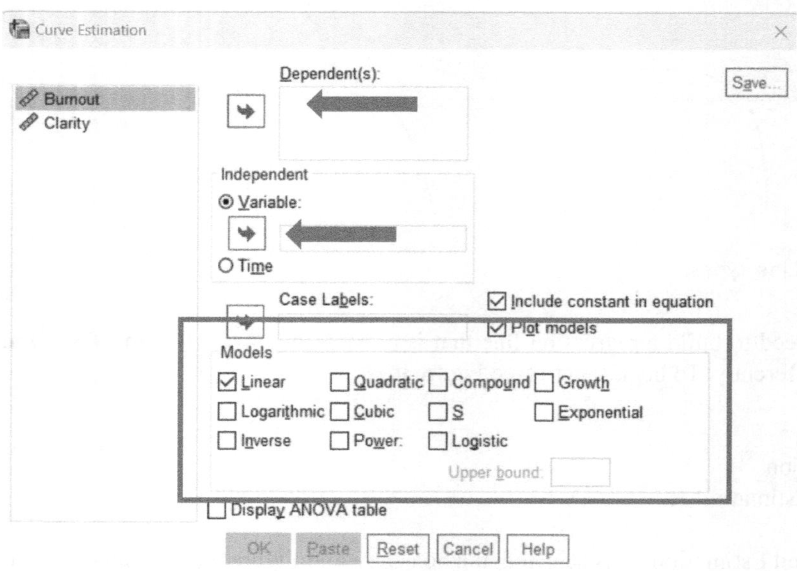

Figure 9.8 Curve options

5. Under Models, click the box next to the type of line that you believe is the best fit to your data
6. Click OK

If you find yourself with curvilinear data, be sure to analyze it with SPSS rather than Excel.

Summary

In this chapter, you've covered the essentials of simple linear regression, a tool that models relationships between two variables to make predictions about how one affects the other. Like Dorian estimating Lego bricks for his walls, you can use simple linear regression to find the best-fit line minimizing the difference between observed and predicted values. In sum, simple linear regression, akin to Dorian's method of estimating Lego bricks, helps make the best predictions with available data, but does not tell you much more than you would learn from correlation. Yet, understanding simple linear regression is a critical step in understanding how to make such predictions when there is more than one independent variable, which you will delve into in Chapter 10.

References

McCroskey, J. C., & Teven, J. J. (1999). Goodwill: A reexamination of the construct and its measurement. *Communications Monographs*, *66*(1), 90–103. https://doi.org/10.1080/03637759909376464

Weber, K., Martin, M. M., & Myers, S. A. (2011). The development and testing of the instructional beliefs model. *Communication Education*, *60*(1), 51–74. https://doi.org/10.1080/03634523.2010.491122

Chapter 10

Multiple Regression

What Will I Learn in This Chapter?

From atop the highest tower of Legoland, Lego Spiderman sits with his notebook, waiting for the next portal into the Spiderverse to open. These portals *seem* to open at random, but he knows that cannot be true. He's been taking notes every day for two weeks, capturing as much data as possible about what is happening when a portal opens:

- What the temperature is
- What time of day it is
- What material (e.g., ground, stone, water) the portal opened on
- How much time has passed since the last opening
- Who is close to the opening

DOI: 10.4324/9781003500117-10

He feels certain that with enough data, he can find a pattern to predict the coordinates in which the next portal will open.

Lego Spiderman is preparing to engage in multiple regression, a statistical analysis trying to predict a dependent variable from the presence of multiple independent variables. In this chapter, you will learn:

- What multiple regression is
- Assumptions of multiple regression
- The types of multiple regression
- How to conduct multiple regression in SPSS

What Is Multiple Regression?

Multiple regression is a way to predict the score of a continuous dependent variable based on the scores of multiple independent variables. Recall from Chapter 9 that the formula for simple linear regression is as follows:

$$Y = \beta X + a.$$

We can easily visualize what this would look like, thanks to our graphing experiences in algebra courses – it will be a straight line.

Multiple regression lines are a bit more complicated. Multiple regression takes the form of:

$$Y = \beta_0 + \beta_1 X_1 + \beta_2 X_2 + \ldots + \beta_n X_{n},$$

where Y is the dependent variable
X's represent the various independent variables
β_0 is the y-intercept
and other betas are the beta weights associated with each variable.

Those beta weights are often called **regression coefficients** when working with multiple regression. The shape of multiple regression lines changes greatly depending on the number of independent variables considered and may take any polynomial shape.

What you have likely guessed from the description of multiple regression so far is there is one difference between simple linear regression and multiple regression: Simple linear regression has only one independent variable, while multiple regression has two or more independent variables. There is no limit to how many independent variables you can utilize in multiple regression.

Test Your Understanding 10.1 – What is the difference between simple linear regression versus multiple regression?

Let us consider an example of simple multiple regression in communication research. Kelly et al. (2023) studied how three supervisor communication variables (i.e., solidarity, humor, and immediate behaviors) affected how much perceived immediacy their subordinates felt they had with their supervisor. This is a test worthy of being studied through multiple regression because the question asked how much three independent variables (i.e., solidarity, humor, and immediate behaviors) affected one dependent variable (i.e., perceived immediacy). Any hypothesis or research question that asks how more than one independent variable affects a continuous dependent variable is appropriate for analysis with multiple regression.

You may be wondering if multiple regression is a better option than a series of correlations. After all, when you got to the end of the simple linear regression chapter, you were told that simple linear regression is no more useful than correlation. Multiple regression is, in fact, a much better option than a series of correlations. The reason for this has to do with statistical power. Recall that for every analysis, you consider the p-value to determine how confident you are that a Type I error has not occurred, and we say that $1 - p$ is the power of the test (how confident we are that we have not found an effect by chance). Correlations have a different p-value for each effect. Multiple regression has a p-value for all effects at once. This means that when $p <.05$ for a multiple regression, you can be at least 95% confident that everything the output predicts about relationships between variables is true, which is far more confident than having only 95% confidence about each individual relationship. Being able to account for multiple independent variables at once is a more powerful test because it is clearer what variance is accounted for by each independent variable and what is **noise** (measurement error), whereas independent correlations are more likely to accidentally credit noise to a relationship between variables.

Let's take a minute to look specifically at the findings of Kelly et al.'s (2023) study. Regression was used to declare how much variance humor, solidarity, and immediate behaviors accounted for in perceived immediacy, but the authors also provide a correlation matrix (p. 11). The correlation matrix for these variables is given in Table 10.1. Each of the correlations in Table 10.1 was statistically significant at $p <.001$, meaning SPSS was over 99% confident that the relationships did not happen by chance per correlation.

However, this did not align with the results of Kelly et al.'s (2023) multiple regression output. Table 10.2 shows the regression output. Notice the relationships are drastically different. Particularly, look at the relationship between humor and perceived immediacy across Tables 10.1 and 10.2. The correlation between perceived immediacy and humor was moderate positive with $p <.001$, but the multiple regression shows a very small negative beta for the same relationship with $p =.73$. This means SPSS was very confident there was a moderate positive relationship between humor and perceived immediacy when looking at the data through correlation, but only

Table 10.1 Correlation Matrix

	1	*2*	*3*
1. Humor	–		
2. Immediate Behaviors	0.56		
3. Perceived Immediacy	0.48	0.65	
4. Solidarity	0.67	0.55	0.59

Table 10.2 Regression Output Regressing onto Perceived Immediacy

	Standardized β	*p-value*
Humor	−0.02	0.73
Immediate Behaviors	0.46	< .001
Solidarity	0.36	< .001

27% confident there was an extremely weak negative relationship between those variables analyzing the same data through multiple regression.

This was not a typo. The output of the analyses was truly that different, even though both were trying to predict how well supervisor humor trended with subordinate perceived immediacy. Having so many independent variables examined at once in the multiple regression gave a much clearer picture of what was happening with the relationship between all the variables, so measurement noise previously credited to the relationship between perceived immediacy and humor in the correlation could be corrected in the regression output. In this paper, Kelly et al. (2023) concluded that the data indicated no relationship between supervisor humor and subordinates' perceived immediacy with their supervisor because the much more powerful multiple regression analysis provided a far more trustworthy overview of the data than correlation, thanks to having much more statistical power.

This is a prime example of how multiple regression is stronger than a series of correlations. The ability to better account for relationships and separate that information from measurement noise is why multiple regression is superior. The more powerful an analysis, the more you can trust it!

Assumptions

Like simple linear regression, multiple regression has the following assumptions:

1. Linearity
2. Independence of errors
3. Homoscedasticity
4. Normality of errors
5. No multicollinearity

These assumptions should seem very familiar to you, having reviewed the same assumptions in simple linear regression.

Because this is a more complicated analysis, some of the methods for checking the assumptions are also more complicated. Linearity and homoscedasticity can still be checked by scatterplot, but this time, a scatterplot will be needed per independent variable, so it can be verified that each independent variable has a linear relationship with the dependent variable. Normality of errors can still be checked with a histogram of the errors. (Later in this chapter, you will see how to ask SPSS to generate a histogram of errors.)

Independence of errors requires an additional step in multiple regression: a **Durbin–Watson test**. Recall that independence of errors is the assumption that an error at one point in time or space is unrelated to an error in another time or space. In social sciences, you almost exclusively see this issue when data is collected at multiple points in time, but it's always best to check. The Durbin–Watson test is used to assess the independence of errors. The output ranges from 0 to 4. The Durbin–Watson scores are ideal at 2. That indicates no **autocorrelation**, correlation across time or space, at all. The closer the Durbin–Watson is to 0, the stronger the positive autocorrelation. The closer to 4, the stronger the negative autocorrelation. This means you want a Durbin–Watson close to 2. Generally, you can consider your data free of influential autocorrelation if your Durbin–Watson score is between 1.5 and 2.5.

Multicollinearity must also be tested differently, through the **Variance Inflation Factor** (VIF). Recall that multicollinearity is what occurs when two or more independent variables are highly correlated with each other to the extent that there is so much measurement overlap that it appears at least two of the assessments are assessing the same variable when they are supposed to be assessing separate variables. VIF tells you whether your regression line fits better than it should because you have two or more independent variables in your line that are measuring the same thing. VIF results range from 1 to infinity. A VIF of 1–5 is considered acceptable. A VIF between 5 and 10 is concerning and should be addressed as a limitation of your model in the write-up of results. A VIF greater than 10 means the multicollinearity is so high that the model should not be trusted, and your independent variable measures need to be reassessed.

Test Your Understanding 10.2 – Why would you run a Durbin–Watson and VIF analysis before interpreting multiple regression results?

Types of Multiple Regression

Building a multiple regression model is much like baking a cake: there are conflicting ideas as to which steps should be completed first. Some bakers think all dry ingredients should be mixed first, then wet ingredients added one at a time. Others think the reverse, that all wet ingredients should be mixed first, then the dry ingredients added slowly, one at a time. Yet others think that all wet ingredients should be combined separately, all dry ingredients should be combined separately, and then the two mixes blended. Each of those options can make an excellent cake. What makes the process correct for the baker is which ingredients and textures they feel are most important to consider first.

There are several options for telling your computer to build a multiple regression model. You may choose:

1. Enter
2. Forward
3. Backward
4. Stepwise
5. Remove

None are always the right choice or always the wrong choice. What you choose will be guided by what you think makes most sense for your data, and truthfully, by how much computing power you have at your disposal. Each of these methods has its own strengths and weaknesses.

The **Enter** method puts all independent variables in the model at once. Your final output will give you a model with all statistically significant and non-significant predictors included. The strength of this method is that it is straightforward and ensures all potential predictors are considered. The weakness is that the model can become too complex, showing some measures to be statistically significant even though they account for little to no variance. Also, the more variables included, the higher the risk of multicollinearity. When using an amply powerful computer, this method should be used only when you are completely certain all predictors are relevant and necessary to your model.

The **Forward** method starts with no predictors and adds them one by one according to which independent variables have the strongest correlation with the dependent variable. The Forward method adds dependent variables one at a time until there are no variables with statistically significant correlations left. The strength of this method is that it avoids overcomplicating the model. The weakness is that it may overlook statistically significant predictors that could not be accounted for with the power of correlation alone. This means the final model may be biased by variables that have been given strength by measurement noise. This is a method to choose when you want a simple model that considers only the subset of statistically significant predictors from correlation.

The **Backwards** method starts with all dependent variables in the model and removes them one at a time, starting with those that have the lowest effect, until all statistically significant dependent variables are accounted for in the model. This is a thorough method that is less likely to be biased than the Enter or Forward methods because each variable has a chance, but measures are removed to avoid complicating the model. The weakness is that, because this is a very complex method, it requires more computing power than the other methods and may not be suitable for all machines. (Don't try this on a Chromebook or any other tablet unless you are very confident in your device's fan or have a small dataset.)

The **Stepwise** method is a combination of the Forward and Backward methods. Like Forward, the Stepwise method adds one predictor at a time. Like Backwards, though, it also removes dependent variables that are found to be non-significant predictors once other variables are considered. As this method combines the strengths of Forward and Backward methods, it is a better option than either, but it can be difficult to interpret, given the complexity of model generations. That high respecification rate also risks **overfitting** the model, which means accepting random noise into the model. Stepwise is the best choice when you want to find the balance between the benefits of Forward and Backward.

Finally, the **Remove** method begins with all variables like Enter but then allows you to analyze and remove variables one at a time based on your own logic. For example, imagine you have two competing theories guiding your research. Remove will allow you to test both theories by strategically removing predictors to find the best explanatory fit. Without a rationale for removing items beyond statistical significance, though, this method will not be of use.

Each method has its own advantages. For the type of machine you are likely to use as a graduate student and the experience you likely have with SPSS so far, Enter will be easiest on your processor and brain. Backwards is the most conservative method but may overtax your machine. When you begin working with data, play with each of these options so you can see the differences in output.

<u>Test Your Understanding 10.3</u> – Which multiple regression method are you likely to choose to build models today and why?

Interpreting Multiple Regressions Output

Table 10.3 shows the output from a multiple regression predicting subordinate latent dissent behaviors based on supervisor solidarity, immediate behaviors, and followership behaviors. This analysis was conducted using the Stepwise method, meaning all three independent variables were initially considered in the model, then removed until only those that were a statistically significant predictor remained.

No matter the method you choose for building your regression model, the final SPSS accepted model is the last one generated. In our case, that is Model 2. With other data, it could be Model 1, Model 20, or Model X – any number depending on how your data relate and the method you choose. So that is your first lesson on interpreting multiple regression results: understand that whatever your last returned model is in the output, that is *the* model – the answer.

The first header in Table 10.3 is Model Summary. Here you have two important pieces of information. First, you have the Adjusted R^2 for your final model. Note this model explains very little about latent dissent, accounting for only 3.5% of the variance. A second important piece of

Table 10.3 Regression Output

Model Summary[c]

Model	*R*	*R Square*	*Adjusted R Square*	*Std. Error of the Estimate*	*Durbin–Watson*
1	.157[a]	0.025	0.021	1.35028	
2	.208[b]	0.043	0.035	1.34025	1.955

a. Predictors: (Constant), Solidarity
b. Predictors: (Constant), Solidarity, ImmediateBehaviors
c. Dependent Variable: LatentDissent

ANOVA[a]

Model		*Sum of Squares*	*df*	*Mean Square*	*F*	*Sig.*
1	Regression	10.832	1	10.832	5.941	.016[b]
	Residual	426.641	234	1.823		
	Total	437.473	235			
2	Regression	18.941	2	9.471	5.272	.006[c]
	Residual	418.532	233	1.796		
	Total	437.473	235			

a. Dependent Variable: LatentDissent
b. Predictors: (Constant), Solidarity
c. Predictors: (Constant), Solidarity, ImmediateBehaviors

Coefficients[a]

		Unstandardized Coefficients		*Standardized Coefficients*			*Collinearity Statistics*	
Model		*B*	*Std. Error*	*Beta*	*t*	*Sig.*	*Tolerance*	*VIF*
1	(Constant)	2.919	0.267		10.929	0.000		
	Solidarity	0.149	0.016	0.157	2.437	0.016	1.000	1.000
2	(Constant)	3.723	0.462		8.056	0.000		
	Solidarity	0.233	0.072	0.246	3.218	0.001	0.701	1.426
	Immediate Behaviors	–0.226	0.106	–0.163	–2.125	0.035	0.701	1.426

a. Dependent Variable: LatentDissent

Excluded Variables[a]

						Collinearity Statistics		
Model		*Beta In*	*t*	*Sig.*	*Partial Correlation*	*Tolerance*	*VIF*	*Minimum Tolerance*
1	Immediate Behaviors	–.163[b]	–2.125	0.035	–0.138	0.701	1.426	0.701
	Followership	–.066[b]	–0.879	0.380	–0.057	0.731	1.368	0.731
2	Followership	–.063[c]	–0.842	0.401	–0.055	0.731	1.369	0.562

a. Dependent Variable: LatentDissent
b. Predictors in the Model: (Constant), Solidarity
c. Predictors in the Model: (Constant), Solidarity, ImmediateBehaviors

information to look at under Model Summary is the Durbin–Watson statistic. Here, the statistics is 1.955, very close to 2, meaning this data shows no risk of violating the assumption of independence of errors.

Next, consider the data under the subheading Coefficients. This gives you the beta weights of your statistically significant predictors and their significance level. Note the two statistically significant independent variables in this model are solidarity ($\beta = .25$, $p = .001$) and immediate behaviors ($\beta = -.16$, $p = .035$). As with simple linear regression in Chapter 10, positive beta weights are indicative of positive relationships between the variables, while negative beta weights are indicative of negative relationships between the variables. The size of the beta weights is interpreted the same as correlation magnitudes, as described in Chapter 8. Thus, you interpret a small positive and significant relationship between solidarity and latent dissent, while there is a small negative and significant relationship between immediate behaviors and latent dissent.

Also, under the Coefficients sub-header are the VIF values for the statistically significant independent variables. The VIF for both predictors is 1.426. This is well under the heuristic of 5, indicating no risk of multicollinearity.

Finally, notice that there are also two models under the Excluded Variables subheading. Again, the last model is the final model, meaning the final model gives you the statistical significance and beta weight for all independent variables that turned out to not be statistically significant predictors of the dependent variable. Here you see that followership was excluded from the model ($\beta = -.06$, $p = .40$).

From this output, you can also build the regression model using unstandardized beta weights. Notice Model 2 under the Coefficients subheading provides unstandardized betas and a constant (y-intercept). So, the model, would be written as:

$$Y = 3.72 - .23X_{\text{Immediate Behaviors}} + .23X_{\text{Solidarity}}.$$

Current research methods in communication do not lend themselves to the need for graphic depictions of such a model. However, other social science fields, such as Economics, do use such models. In these fields, you are likely to see the equation written as follows:

$$Y = 3.72 - .23X_{\text{Immediate Behaviors}} + .23X_{\text{Solidarity}} + \varepsilon$$

Again, the Greek letter ε represents the error in the model because this is our line of *best* fit, not an exact fit to the data. If you happen to be an economics student or studying an allied field, notice that each statistically significant beta weight has its own error to contribute to the overall model, and is estimated under the Model Summary heading.

Lastly, take a moment to look at the F-value in the ANOVA table. To reiterate, the F-value tells you whether the variance explained by your dependent variables is greater than the variance still unexplained. A statistically significant F-value tells you the model is too weak to be trusted, meaning there is more noise than variance explained. In multiple regression, it is possible to have statistically significant beta weights and still have an F-value that is not statistically significant. However, this is only the case if you use the Enter method of building your model, though. If you use the Enter method, it is important to look at the F-value and see if the volume of non-significant dependent variables is causing too much noise in your model for it to be useful. If so, the regression should be run again with only the variables that yielded statistically significant beta weights considered as dependent variables. Because this example used a stepwise method to build the model, it is unsurprising that the F-value is statistically significant at $p = .006$ for the second model.

Reporting Regression Output

In Chapter 9, results of a simple linear regression were presented as simply the beta value and *p*-value. That is common for simple linear regression because that analysis does not give you any more information than a correlation analysis; it is simply another way to think about the data. Readers expect more when you are reporting a multiple regression, though.

First, because models with multiple dependent variables are much more complex, the reader needs to know that assumptions were checked. It is always important to start by saying which type of regression was used. Readers will want to know both the Durbin–Watson and the VIF values. Readers will also want to know the beta weights and *p*-values in the supported model as well as those of dependent variables that were not retained in the supported model. Finally, if the Enter method was used to develop the regression model, readers will want to know the *F*-value. (Please don't use the Enter method. It is the weakest option and requires more work in the write-up.) Here is an example of how the authors might write up the results of a multiple regression to see if immediate behaviors, solidarity, and followership explain latent dissent:

> The hypothesis was tested through multiple regression using the stepwise method. Assumptions of appropriateness of the data for regression analysis were supported through VIF = 1.43 and Durbin–Watson = 1.96. Model 2 was the final model, which supported both solidarity (β =.25, p =.001) and immediate behaviors (β = −.16, p =.04) as statistically significant predictors of latent dissent. Followership (β = −.06, p =.40) was not a statistically significant predictor of latent dissent.

Keep in mind not every regression model will have statistically significant results. Sometimes the data just does not work as expected. In these instances, still report the VIF and Durbin–Watson, then state that none of the dependent variables were statistically significant predictors before providing all beta weights and *p*-values.

Dummy Coding

At some point in time, you may find yourself in a position in which you want to know how well a categorical variable predicts an independent variable in a multiple regression. In our introduction, Lego Spiderman has such a problem because he wants to incorporate the type of material each portal appears in. Multiple regression can be run with categorical variables as predictors!

To do this though, you must do what is called **dummy coding** your data, a process of creating new binomial columns of data for each categorical variable option. Imagine you have data on participant gender that you wish to use as an independent variable in a multiple regression. Imagine that Table 10.4 represents the gender variable in your first 10 rows of data where male is coded as 1, female is coded as 2, and third gender responses are coded as 3. You cannot use your data as is in the regression model because SPSS and Excel are only as smart as your ability to operate them, and therefore those programs will assume this is interval data. SPSS may try to stop you from using this variable as interval data if you have it set as nominal in your Variable View, but you can override that warning if you wish and run any nonsense analysis with gender as a continuous variable you wish. You need to be smarter though and properly prepare your data.

Recall from Chapter 8 that a correlation of 0 means no relationship at all, and a correlation of 1 means a perfect relationship (i.e., these variables are so perfectly aligned that they must be the same thing). Using this logic, you can dummy code your nominal variables, meaning you will create a new column of data for each category of response and assign a total match (1) or no match (0). To make this less abstract, look at Table 10.5.

Table 10.4 Sample Gender Data

Gender
1
2
3
1
1
2
2
2
1
1

Table 10.5 Dummy-Coded Data

Gender	*Male*	*Female*	*Third Gender*
1	**1**	0	0
2	0	**1**	0
3	0	0	**1**
1	**1**	0	0
1	**1**	0	0
2	0	**1**	0
2	0	**1**	0
2	0	**1**	0
1	**1**	0	0
1	**1**	0	0

Look first at the Male column. Note that for each row in which the Gender column indicated that the participant was a male, the Male column has a 1. Likewise, for each row in the Gender column indicating that the participant was not male, the Male column has a 0. As such, 0 and 1 are used to show whether gender was a $r = 0.00$ or $r = 1.00$ match to being male. This same pattern is seen in the Female and Third Gender columns in which 1 indicates a match and 0 indicates no match. Now, in lieu of using the Gender variable as an independent variable in the regression model, Male, Female, and Third Gender can each be used as independent variables to see if the presence of those genders are statistically significant predictors of the dependent variable.

Interpreting the results of dummy-coded data analyses involves the statistical significance and the polarity of the data. Imagine, for example, that Table 10.6 contains the standardized beta weights and *p*-values of your multiple regression output for the dummy-coded gender data. Third gender has a *p*-value greater than .05, so a participant identifying as Third Gender would not be a statistically significant predictor of your dependent variable. Male and female both have *p*-values lower than .05, though, and as such are statistically significant predictors of the dependent variable.

The male beta weight is negative. That means male participants are likely to have lower values of the dependent variable. Conversely, female beta weight is positive. That means female participants are likely to have higher values of the dependent variable. If you run your multiple regression with only the dummy variables, the Adjusted R^2 will indicate what percent of variance in your dependent variable is accounted for by participant gender alone.

Table 10.6 Sample Dummy Code Output

	Standardized β	*p-value*
Male	−0.20	.04
Female	0.31	.01
Third Gender	0.12	.21

Multiple Regression in SPSS

Back in Chapter 3, it was previewed that eventually the book would stop showing you how to complete analyses in Excel. This is that moment. While Excel can perform multiple regression, the data setup is particular and finicky, the computing power is greater than required by SPSS for the same analyses, and the output is not organized as well as in SPSS. Thus, for the sake of making sure the statistical concepts remain clear and are not lost in managing the computer programs, here the book transitions to teaching multiple regression in only SPSS.

To begin computing a multiple regression in SPSS, follow the same steps you used to set up the linear regression, as shown in Figure 10.1:

1. Click Analyze
2. Click Regression
3. Click Linear

As with linear regression, those steps take you to the linear regression menu. As seen in Chapter 9, you will use the arrows and boxes shown in Figure 10.2 to move independent and dependent variables into the model:

4. Click on the dependent variable, and then use the arrow beside Dependent to move that variable into the dependent variable box
5. Click on each of the independent variables and use the arrow beside Model 1 to move those variables into the Model 1 box

From this point forward, there are additional steps required that you did not practice in Chapter 9.

6. The first new step is to click on the dropdown menu beside Method and select the method that you would like to use to build your regression model. The default is always set to Enter because that is the method that will require the least computing power. You should choose the method that best meets the needs of your model (or that your machine has the processing power to manage).
7. Then click on the button labeled Statistics . . .

The Statistics. . . button will produce the menu shown in Figure 10.3. Here is where you begin telling SPSS to test your assumptions.

8. Under Residuals, click the box next to Durbin–Watson to produce the Durbin–Watson test of independence of errors
9. On the right side of the Statistics . . . box, click on the box next to Collinearity diagnostics to calculate VIF, testing for multicollinearity
10. Click Continue

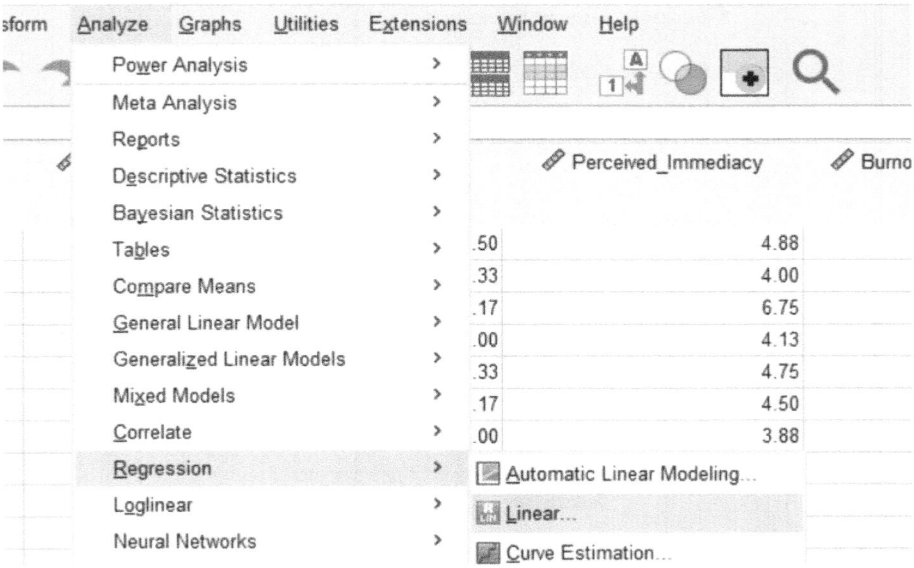

Figure 10.1 Regression menu.

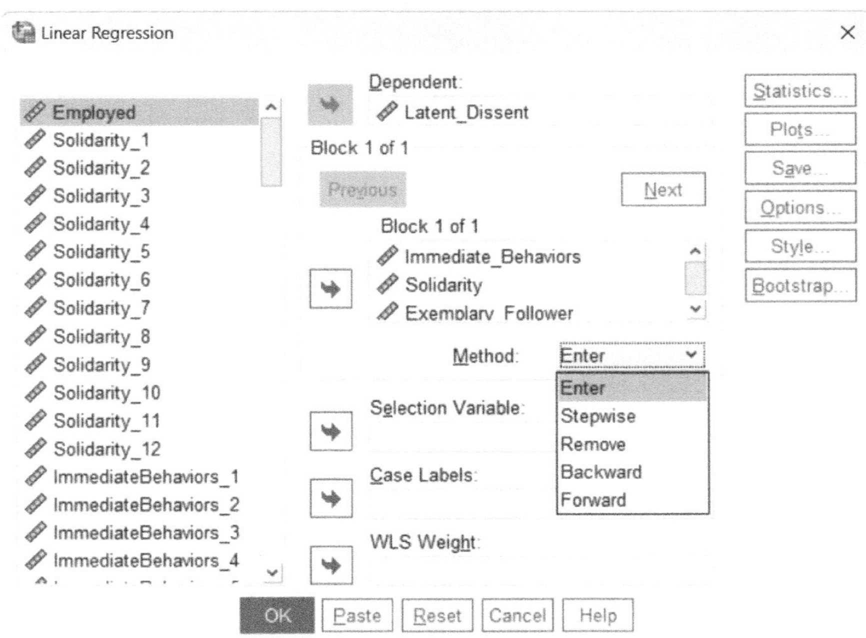

Figure 10.2 Regression method.

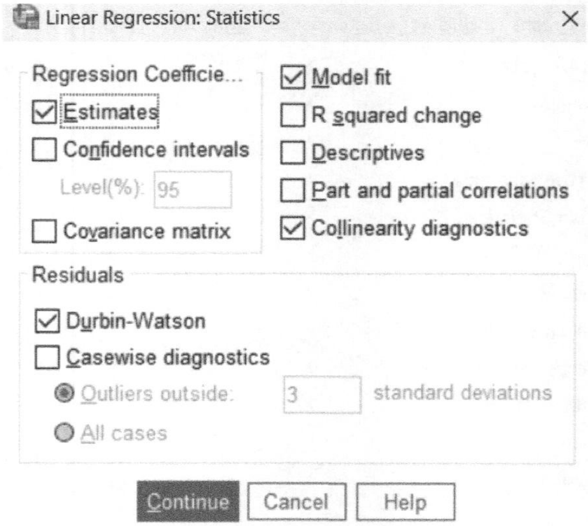

Figure 10.3 Statistics.

Figure 10.4 Regression plots.

This will return you to the regression menu shown in Figure 10.2. The final step is to have SPSS automatically create a histogram of the distribution of errors so that you can visually confirm that the independence of errors assumption is met.

11. Click Plots . . .
12. This will produce the menu shown in Figure 10.4. On the bottom left of this menu, under Standard Residual plots, click the box next to Histogram
13. Click Continue
14. Click OK in the regression menu

Once these steps are completed, your multiple regression results will be printed in the Output window. You will notice that there is more output provided for you than previewed in this chapter. Some of this information was computed in the process of producing the output you have read about in this chapter. Other pieces of that information give you different details about your analysis. The purpose of this book is to introduce you to what a social scientist, particularly in the field of communication, *needs* to know. If multiple regression becomes a method you use frequently in your research, you will benefit from understanding all the associated output. If you want to know more about the additional output, Howell (2013) is an excellent reference.

Summary

This chapter explored multiple regression; the analysis appropriate for predicting a continuous variable when there are multiple independent variables. The assumptions of what your data look like to run a multiple regression were overviewed, as well as the types of methods for having a computer program build the model. You also learned how to run multiple regressions in SPSS. Much like Lego Spiderman, who meticulously collected data to predict portal openings, you've learned how to uncover patterns to make predictions in a multivariate dataset.

References

Howell, D. C. (2013). *Statistical methods for psychology*. Wadsworth Cengage Learning.

Kelly, S., Zeng, C., & Cundall Jr., M. K. (2023). Subordinate articulated dissent as influenced by supervisor behaviors: The hazards of humor. *International Journal of Business Communication, Early Access*. https://doi.org/10.1177/23294884231166405

Chapter 11

t-Tests

What Will I Learn in This Chapter?

The classic battle against good and evil – Darth Vader, master of the Dark Side, against Luke Skywalker, young Jedi and master of the Light Side of the Force. On multiple occasions, these two titans of the Force have battled one another. In the original Star Wars trilogy (*Star Wars: A New Hope, Empire Strikes Back,* and *Return of the Jedi*), Vader and Luke fought twice. Their first battle was on Cloud City, and Luke lost this one. (Spoiler alert!) Their second battle was on the Death Star, and Luke won this one, with Vader dying shortly after the battle. With these two titans being 1–1 in their battles, who is the stronger Force wielder? This is a debate among Star Wars nerds, we mean fans. If it were possible to calculate Vader's and Luke's force capability, and give each capability a mean score, and then compare the two, the difference between these two could be statistically compared. This comparison is one kind of a *t*-test.

DOI: 10.4324/9781003500117-11

In this chapter, you will be introduced to how we statistically compare two means. Specifically, in this chapter you will:

- Understand what *t*-tests are
- Explore the two primary kinds of *t*-tests
- Learn how to run *t*-tests in Excel and SPSS

What Are t-Tests?

In Chapters 8–10, you have read about how variables are related to one another. These chapters explored how statistical tests can show correlation (association) and how you can use statistical tools to predict Y from X (regression) or Y from a series of variables (multiple regression). Sometimes researchers are not interested in exploring relationships but want to explore the extent to which variables differ among groups. Here are a few examples. First, let's say you want to understand the frequency of which adolescents and older adults differ in online news consumption. You could measure online news consumption among adolescents and older adults for one week, and then statistically compare their viewership. Second, let's say you're interested in the effects of a new migraine medication. You could give one group the medication and another a placebo and see the difference between the two groups. Third, you want to know if students in a public speaking class have decreased levels of anxiety after taking the course. You can measure their level of speech anxiety at the start and end of the course.

Each of these scenarios examines differences in a mean score and can be statistically explored using a kind of *t*-test. The first two scenarios involve exposing people or comparing people from different conditions (independent design or between-group), while the third scenario involves a single group tested at different points in time (paired sample or dependent). There are two primary kinds of *t*-tests, which will be discussed further at length later, the independent samples and the dependent samples *t*-test. Both parametric tests have similar rationale/assumptions (Field, 2009; Hayes, 2005).

1. Variable – The dependent variable must be at least at the interval level. The independent variable must be nominal or ordinal level. Refer to Chapter 2 for a discussion on levels of measurement.
2. Normality – Both samples are approximately normally distributed. Refer to Chapter 6 for a discussion on normality.
3. Representative Sampling – The data should come from representative samples. Refer to Chapter 2 for a discussion on sampling.
4. Homogeneity of Variance – Both samples should have equal or similar levels of variance.
5. The larger the sample, the less likely the difference between the means analyzed is created by sampling error.
6. The further apart the means, the less likely the difference was created by sampling error.

<u>**Test Your Understanding 11.1**</u> – What are the six assumptions of *t*-tests?

The Two Primary Kinds of t-Tests

There are two primary kinds of *t*-tests: the dependent samples *t*-test and the independent samples *t*-test. A **dependent samples *t*-test** (sometimes called paired samples) is a *t*-test used when there are two conditions (data collection points), and the same participants take part in both conditions.

An **independent samples *t*-test** (sometimes called independent measures) is a *t*-test used when there are two conditions, and different participants take part in both conditions.

Test Your Understanding 11.2 – Which *t*-test would you use when you have two conditions and the same participants?

Dependent Samples t-Test

Dependent samples *t*-tests compare the means of two related groups. Often, this test is used to compare mean scores at one point in time to the mean scores at another point in time. For example, one of the writers of this text regularly gives students a test on the first day (pre-test) of Statistics 101 and the same test on the last day (post-test). The mean results are then compared to see if students improved in their statistical understanding. Previous research in communication has utilized dependent samples *t*-tests to answer communication questions. For example, Mohammadi et al. (2023) found that English as a Foreign Language (EFL) teachers had higher levels of intercultural sensitivity after training. Semlak and Pearson (2008) used a dependent samples *t*-test to show differences in how students perceived teachers of different age groups. Bond et al. (2004), in their analysis of prisons, showed that non-inmates did not have a statistically significant difference from a pre-test to post lie decision test.

Below, we outline the process for carrying out a dependent samples *t*-test mathematically, with an example from 15 Statistics 101 students. (It is recommended to have at least 20 participants per group to run a *t*-test, but this smaller dataset has been used to better fit the pages here.) Don't worry, SPSS and Excel will do all the math for you. However, it is important to know what these programs are doing. Let x = the test score before the course and y = the test score after the course. To test the null hypothesis that there is no difference in test scores over the course of the semester, you would follow this procedure:

1. Compute the difference ($d_i = y_i - xi$) between the two testing points (observations).
2. Calculate the mean difference, $\bar{d}$.
3. Compute the standard deviation of the differences, *sd*, and from this, calculate the standard error of the mean difference, $SE(\bar{d}) = \frac{s_d}{\sqrt{n}}$.
4. Compute the *t*-statistic: $t = \frac{\bar{d}}{SE(\bar{d})}$. Assume your $df = n-1$.
5. Use the *t*-distribution table in Appendix T to compare your *t*-value and determine the *p*-value for the test.

The following is what each statistical symbol means above:

- $\bar{d}$ = the difference in means
- y_i = score after the course
- x_i = score at start of course
- s_d = standard deviation of the difference
- SE = standard error
- n = sample size
- t = *t*-statistic
- r = correlation or effect size

Table 11.1 Statistics Test Pre and Post

Student	*Day One Test Result*	*Last Day Test Result*	*Difference*
1	55	62	7
2	47	76	29
3	25	45	20
4	67	91	24
5	70	71	1
6	75	74	−1
7	55	65	10
8	54	64	10
9	68	76	8
10	53	63	10
11	37	74	37
12	81	80	−1
13	48	79	31
14	27	61	34
15	46	78	32

The mean difference $\bar{d}$ = 251/15 = 16.66 and the standard deviation = 13.41. Thus, the $SE(\bar{d})$ $= \frac{s_d}{\sqrt{n}} = \frac{13.41}{\sqrt{15}}$ or 3.46. So, for the *t*-value, we have $t = \frac{\bar{d}}{SE(\bar{d})}$ or $t = \frac{16.66}{3.46} = 4.82$, with 14 degrees of freedom (*df*). Looking at the critical values of *t* for two-tailed tests in Appendix T, we can see that $p <.001$. Thus, there is a significant difference in Statistics 101 student tests from the start (pre-test) to the end of the semester (post-test). In addition to reporting the *t*-value, when running a dependent samples or paired samples *t*-test, you should also be interested in the effect size. Essentially, you should look at the magnitude (strength) of the relationship between your constructs. Again, SPSS or Excel will tell you the effect size. However, if you are interested in the equation for the effect, here it is below:

$$r = \sqrt{\frac{t^2}{t^2 + df}}$$

In terms of the statistics students, the $r = .58$. The overall result could be written up as such: A dependent samples *t*-test was conducted on Statistics 101 students' statistical knowledge at the start and end of the semester. The results reveal a significant improvement in students' understanding of statistics from the start ($M = 53.87$, $SD = 16.48$) to the end ($M = 70.60$, $SD = 10.91$) of the semester, $t(14) = 4.82$, $r = .58$, $p <.001$.

Test Your Understanding 11.3 – Why is reporting the effect size important with dependent samples *t*-tests?

Independent Samples t-Test

Independent samples *t*-tests compare mean scores of two unrelated groups. This test is often used to compare means of participants from two distinct populations. Aside from the six assumptions listed earlier for *t*-tests, there is one additional assumption for an independent samples *t*-test:

1. Independence – Observations or data from one sample are independent of observations or data from the other sample. Refer to Chapter 6 for a discussion on Independence.

In terms of independent *t*-tests, Ashwell et al. (2022) used an independent samples *t*-test to examine confidence in the COVID-19 vaccine among those individuals who identify as able and unable to receive a COVID-19 vaccination. The researchers found that those individuals who are medically able to be vaccinated have higher confidence than those who identify as medically unable to be vaccinated. The analysis was conducted among two independent populations (groups): those who identify as medically able to receive the COVID-19 vaccine, and those who identify as medically unable to be vaccinated.

Below, we outline the process for carrying out an independent samples *t*-test mathematically, with an example from 30 Statistics students' final exam results (15 first-year and 15 second-year). Don't worry, again, SPSS and Excel will do all of this for you. However, it is important to know what these programs are doing.

1. Compute the difference between the two sample means, $\overline{x}_1 - \overline{x}_2$.
2. There are two equations for computing an independent samples *t*-test. The equation differs based on whether equal variances are assumed (homogeneity). If the two independent samples are assumed to have come from populations with similar population variances, the *t*-test equation is as follows:

$$t = \frac{\overline{x}1 - \overline{x}2}{s_p\sqrt{\frac{1}{n_1} + \frac{1}{n_2}}}$$

In this equation,

$$s_p = \sqrt{\frac{(n_1 - 1)s_1^2 + (n_2 - 1)s_2^2}{n_1 + n_2 - 2}}$$

When equal variances are not assumed, the equation is as follows:

$$t = \frac{\overline{x}1 - \overline{x}2}{\sqrt{\frac{s_1^2}{n_1} + \frac{s_2^2}{n_2}}}$$

Table 11.2 Sample Data

Group	*Sample Size*	*Sample Mean*	*Sample Standard Deviation*
1st Year	15	71.73	10.05
2nd Year	15	79.20	7.96

In Chapter 6, we talked about homogeneity of variance and how it relates to parametric data. Parametric tests, such as correlations, regression, ANOVA, and *t*-tests, assume that the variances between groups are approximately equal. However, this is not always the case. Thus, to test for variations in homogeneity, we use Levene's test. **Levene's test** tests the assumption that variances in groups are equal. If the test is significant, $p < .05$, the null hypothesis is incorrect, and the variances are statistically different. Thus, homogeneity of variance has been violated. When we run independent samples *t*-tests in SPSS, it automatically runs this test for us. We will show how to report this result shortly.

The following is what each statistical symbol means in the above equations:

- $\bar{x}_1$ = the mean of the first sample
- $\bar{x}_2$ = the mean of the second sample
- n_1 = the sample size of the first sample
- n_2 = the sample size of the second sample
- s_1 = the standard deviation of the first sample
- s_2 = the standard deviation of the second sample
- s_p = the pooled standard deviation

When solving by hand, use the *t*-distribution table in Appendix T to compare your *t*-value to the *t*-distribution and determine the *p*-value for the test.

The overall result could be written as such: An independent samples *t*-test was conducted on Statistics 101 students' statistical knowledge, comparing first-year ($M = 71.73$, $SD = 10.05$) and second-year ($M = 79.20$, $SD = 7.96$) students' final exam scores. Results reveal second-year students scored significantly higher, $t(28) = 2.26$, $p = .03$.

Test Your Understanding 11.4 – Why is Levene's Test important in conducting an independent samples *t*-test?

Summary Thoughts on t-Tests

A few thoughts before we go into how to conduct *t*-tests. First, it is important to consider your sample size when conducting dependent and/or independent samples *t*-tests. Think about what you've learned about normality, reliability, and the Central Limit Theorem so far. If the sample size is too small, then the power of the *t*-test will more than likely be too small to detect meaningful results in the data. You also cannot achieve a normal distribution with less than 20 data points. This is why we recommended a minimum of 20 data points per cell earlier. When assumptions of *t*-tests are violated, the accuracy of results can be compromised, and so it's critical to ensure that all assumptions are met before making decisions based on your results.

How to Conduct t-Tests in Excel and SPSS

Dependent Samples t-Test in Excel

For this chapter, two sets of data will be analyzed. The first will help clarify how to conduct a dependent samples *t*-test. This data set explores speech anxiety at the start and end of the semester for 20 students. 20 students were given a measure of anxiety, assessing their level of anxiety ranging from 0 to 100 at the start of a public speaking class (Day 1 of the semester) and on the final day of the semester.

Okay, we know we just told you that we would quit using Excel in the last chapter. We really are going to quit showing you how to run analyses in Excel. However, because the steps to conduct a t-test are straightforward in Excel, we will provide you with those:

1. Go to the Data Analysis menu
2. Select t-Test: Paired Two Sample Means
3. OK
4. Provide the cell range for data from the first time or first point in "Variable 1 Range"
5. Provide the cell range for data from the second time or second point in "Variable 2 Range"
6. Click OK

With these selections, Excel will provide your output on a new sheet.

Dependent Samples t-Test in SPSS

Running a dependent samples t-test in SPSS is relatively quick. First, as shown in Figure 11.1, open the paired samples t-test menu by clicking:

1. Analyze
2. Compare Means and Proportions
3. Paired Samples t-Test

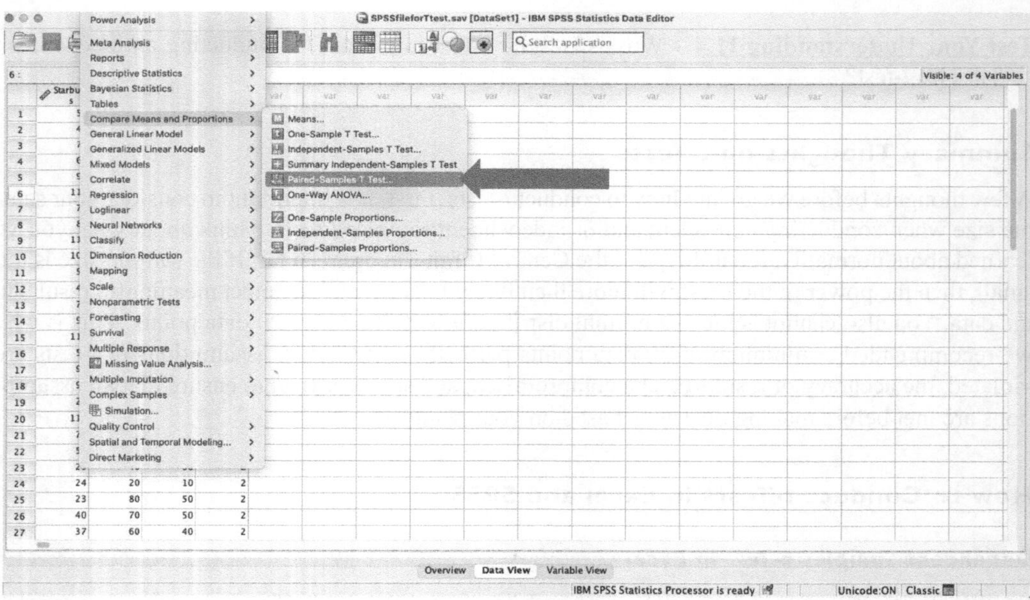

Figure 11.1 Calling up the dependent/paired samples *t*-test in SPSS.

These steps will open the window for the paired samples (dependent samples) *t*-test shown in Figure 11.2. Within this window:

4. Click on the first variable you want to analyze and then the arrow to move it to "Variable 1." Then click on the second variable you want to analyze and then the arrow to move it to "Variable 2." Then click "OK."

You will get the following outputs, which are important for you to have when writing up your results. Table 11.3 is the Paired Samples Statistics Output. This output shows the means, standard deviations, sample size, and standard errors for your two samples.

The next output, Table 11.4, is the Paired Samples Correlations. As discussed earlier, this output includes the effect size, or correlation between the two samples. For your ease, we have highlighted it.

The third output, Table 11.5, is the Paired Samples Test. This is where you will see the mean difference (15.60), the standard deviation of the difference (11.256), the SE of the difference (1.78), the *t*-statistic (8.765), the degrees of freedom/*df* (39), and the level of significance (p). Let's assume we hypothesized anxiety would decrease; this is a one-sided hypothesis (p <.001).

In writing up these results, you would write the following: It was proposed that speech anxiety would decrease from the start (M = 61, SD = 16.34) to the end (M = 45.40, SD = 18.41) of the semester. This hypothesis was supported, $t(39)$ = 8.77, r =.80, p <.001. You will notice that each value has been rounded up to the second decimal point.

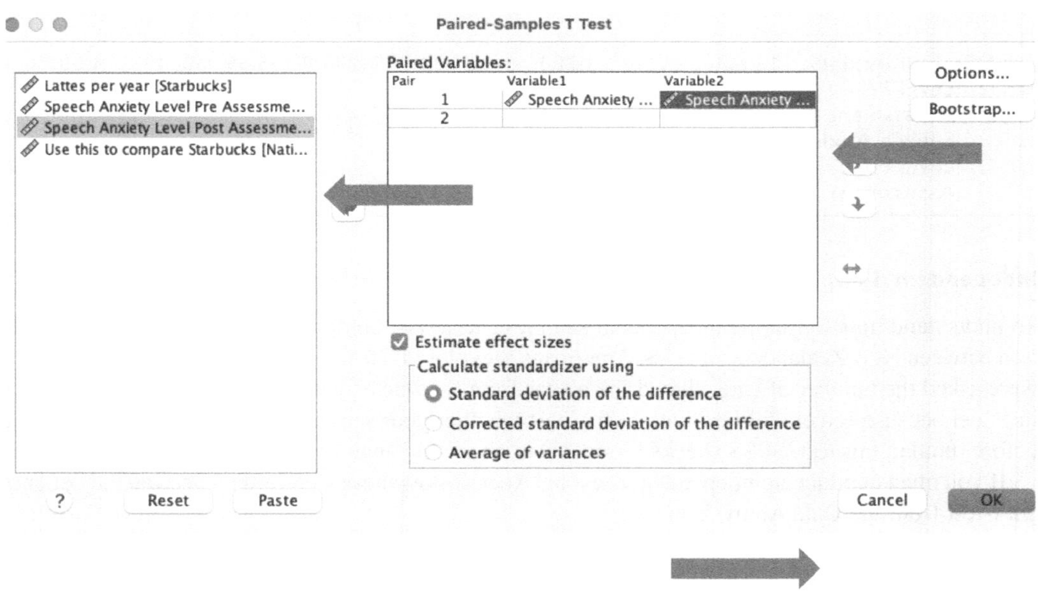

Figure 11.2 Dependent/paired samples *t*-test box.

Table 11.3 Paired Samples Statistics

		Mean	N	Std. Deviation	Std. Error Mean
Pair 1	Speech Anxiety Level Pre Assessment	61.00	40	16.336	2.583
	Speech Anxiety Level Post Assessment	45.40	40	18.405	2.910

Table 11.4 Paired Samples Correlations

		N	Correlation	Significance	
				One-Sided p	Two-Sided p
Pair 1	Speech Anxiety Level Pre Assessment & Speech Anxiety Level Post Assessment	40	.796	<.001	<.001

Table 11.5 Paired Samples Test

		Paired Differences							Significance	
		Mean	Std. Deviation	Std. Error Mean	95% Confidence Interval of the Difference				One-Sided p	Two-Sided p
					Lower	Upper	t	df		
Pair1	Speech Anxiety Level Pre Assessment – Speech Anxiety Level Post Assessment	15.600	11.256	1.780	12.000	19.200	8.765	39	<.001	<.001

Independent Samples t-Test in Excel

To understand how to run the independent samples *t*-test, data comparing Starbucks latté consumption between New Zealanders and U.S. Americans is explored. 20 New Zealanders and 20 Americans were asked the number of lattés they drink a year from Starbucks. Independent *t*-tests are a problem in Excel because Excel will need you to first test whether your sample has equal variances by hand before running this test. SPSS checks for equal variances and analyzes accordingly for you.

If you must conduct an independent *t*-test in Excel, follow these steps after selecting the dependent t-test from the Data Analysis menu:

1. Select the ranges for your dependent variable associated with the first grouping variable, placing that in the Variable 1 Range
2. Select the ranges for your dependent variable associated with the second grouping variable, placing that in the Variable 1 Range
3. Click OK

Again, these steps will move your output to a new sheet.

Independent Samples t-Test in SPSS

Running an independent samples *t*-test is also relatively simple in SPSS. First, as shown in Figure 11.3, open the independent samples *t*-test menu by clicking:

1. Analyze
2. Compare Means and Proportions
3. Independent Samples *t*-Test
4. Click on the interval/ratio level dependent variable you want to analyze and then the arrow to move it to "Test Variable(s)." In this case, the variable is "Lattés per year"
5. Then click on the Grouping variable (independent variable) you are using. For example, (male/female) or (red/blue). In this case it is "Nation." Click on it and move it to "Grouping variable"
6. You will then need to define groups. Click on "Define groups." In this case, the only two options are "1" and "2." "1" = the United States and "2" = New Zealand. Click "Continue." Choosing these two options tells SPSS to compare the U.S. and New Zealand on the average number of lattés from Starbucks consumed per year. Then click "OK"

You will get the following outputs, which are important for you to have when writing up your results. Table 11.6 is the Group Statistics output. This shows you the means, standard deviations, *N* (sample size) per group, and standard errors for each sample.

Table 11.7 is the Independent Samples Test. This output shows Levene's Test for Equality of Variances, the *t*-statistic, the degrees of freedom/*df,* the level of significance (*p*), and the standard error of the difference.

As shown in Table 11.7, Levene's test is significant, which means equal variances cannot be assumed. Thus, we will interpret our *t*-test using the lower line of this table. Let's assume we said

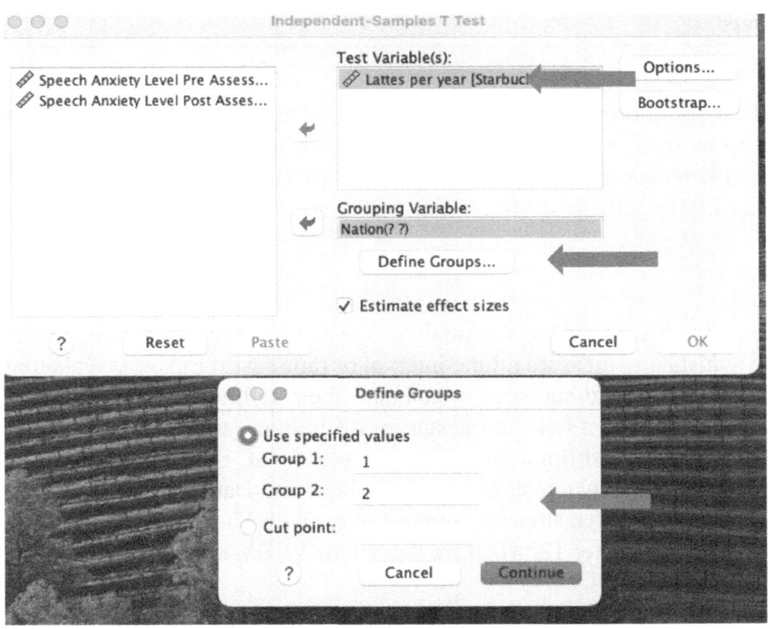

Figure 11.3 Independent samples *t*-test.

Table 11.6 Output

Group Statistics

	Use this to compare Starbucks	*N*	*Mean*	*Std. Deviation*	*Std. Error Mean*
Lattés per year	US	20	80.20	27.031	6.044
	New Zealand	20	21.00	18.539	4.145

Table 11.7 Further Output

Independent Samples Test

		Levene's Test for Equlity of Variances		*t-test for Equality of Means*							
						Significance				*95% Confidence Interval of the Difference*	
		F	*Sig.*	*t*	*df*	*One-Sided p*	*Two-Sided p*	*Mean Difference*	*Std. Error Difference*	*Lower*	*Upper*
Lattés per year	Equal variances assumed	5.091	.030	8.077	38	<.001	<.001	59.200	7.329	44.363	74.037
	Equal variances not assumed			8.077	33.636	<.001	<.001	59.200	7.329	44.299	74.101

New Zealanders would drink fewer Starbucks lattés than Americans. Thus, we would report the result of the test as such: Results of an independent samples *t*-test showed that New Zealanders ($M = 21$, $SD = 18.54$) consumed significantly fewer Starbucks lattés than Americans ($M = 80.20$, $SD = 27.03$), $t(33.64) = 8.08$, $p < .001$.

Summary

In this chapter, you've covered the essentials of *t*-tests, a tool that compares two means. If there were a way to statistically measure (at the interval or ratio level) Luke Skywalker and Darth Vader's force abilities, we could statistically compare their abilities using an independent samples *t*-test. Or if we could measure Luke at the start of his training with Obi-Wan Kenobi and after his training with Yoda, we could run a dependent samples *t*-test. Such analyses could show us how force abilities changed for Luke over time, or how Luke and Darth differ. Adding to the mix, what if we wanted to compare force strength among Luke, Darth Vader, and Yoda? This is a statistical test we will explore in Chapter 12, when we delve into ANOVA.

References

Ashwell, D., Cullinane, J., & Croucher, S. M. (2022). COVID-19 vaccine hesitancy and patient self-advocacy: A statistical analysis of those who can and can't get vaccinated. *BMC Public Health*, *22*, 1296. https://doi.org/10.1186/s12889-022-12661-4

Bond, G. D., Malloy, D. M., Thomspon, L. A., Arias, E. A., & Nunn, S. N. (2004). Post-probe decision making in a prison context. *Communication Monographs*, *71*(3), 269–285. https://doi.org/10.1080/03634520420000288328

Field, A. (2009). *Discovering statistics using SPSS* (3rd ed.). Sage.

Hayes, A. F. (2005). *Statistical methods for communication science*. Lawrence Erlbaum Associates.

Mohammadi, M., Yousefi, M. H., & Salimi, A. (2023). Effects of intercultural sensitivity and critical thinking trainings on teachers' professional development. *Journal of Intercultural Communication Research*, *52*(3), 334–356. https://doi.org/10.1080/17475759.2023.2199299

Semlak, J. L., & Pearson, J. C. (2008). Through the years: An examination of instructor age and misbehavior on perceived teacher credibility. *Communication Research Reports*, *25*(1), 76–85. https://doi.org/10.1080/08824090701831867

Chapter 12

Analysis of Variance (ANOVA)

What Will I Learn in This Chapter?

In Chapter 11, we compared the force capabilities between Luke Skywalker and Darth Vader. For this comparison, an independent samples *t*-test was the most appropriate statistical tool, because we had two people measured on their force abilities (assuming an interval level measure). Above, we have members of the Lego Jedi Council posing for a group photo. After this photo, Lego Jedi Master Ki-Adi-Mundi (he's the one with the tall pink head) jokingly says, "we should compare force abilities between planets." To which Lego Master Mace Windu (the Black Jedi behind Lego Master Yoda's right shoulder) says, "bring it on Mundi, let's see who among the Council is the strongest Jedi." In Chapter 11, a *t*-test was used to compare Jedi to Jedi. To compare the force abilities across all the Lego Jedi Council members at once, we need a more robust statistical test. This is where the analysis of variance (ANOVA) comes into play.

In this chapter, you will learn all about the ANOVA test and how it compares multiple means simultaneously. Specifically, in this chapter, you will:

- Understand the basic principles of an ANOVA
- Explore the *F*-statistic and its importance in ANOVA
- Look at post-hoc procedures
- Learn how to run an ANOVA in Excel and SPSS

DOI: 10.4324/9781003500117-12

Basic Principles of ANOVA

A **one-way ANOVA** is a statistical test used to compare more than two groups on an interval or ratio level variable. For example, Croucher et al. (2024) compared extraversion and argumentativeness levels (interval level variables) between four nations, the United States, Canada, Australia, and New Zealand (nominal/categorical variable). In their analysis they found that participants from the U.S.A. were higher on argumentativeness and extraversion, while New Zealanders scored the lowest on argumentativeness and extraversion.

An ANOVA is a parametric test. Thus, it has many assumptions that should be familiar to you already:

1. Continuous Dependent Variable – The dependent variable must be continuous. This means interval and ratio data are always appropriate to be the dependent variable, but in rare cases, ordinal variables can be a dependent variable in ANOVA.
2. Independent Variable – The independent variable (grouping variable) must be a nominal or ordinal level variable.
3. Normality – All of the samples should be normally distributed.
4. Representative Sampling – Data should come from representative samples.
5. Homogeneity of Variance – All samples should have equal or similar levels of variance. However, if there are deviations, there are corrections that can be conducted, which we will talk about later in this chapter. The greater the differences in group sizes, the greater the impact that homogeneity of variance can have on the results.
6. The larger the sample, the less likely difference(s) in means is attributed to sampling error.
7. The further apart the means, the less likely difference(s) is attributed to sampling error.
8. The more homogenous the sample, the less likely difference(s) is attributed to sampling error.

Test Your Understanding 12.1 – What is an ANOVA used to understand?

The *F*-Statistic or Test

The ***F*-test** is an extension of the *t*-test. This statistical test determines whether the variance between groups is significantly different from the variance within groups when comparing more than two populations or samples. There are a few key things to know about the *F*-test. First, it is an **omnibus test**, meaning it assesses the significance of multiple parameters at once. Second, the results indicate the extent to which group variances differ. (Although not entirely accurate, you can think of it as checking whether means differ across any groups.) However, the *F*-test does not identify which specific groups are different; it only tells you that at least two groups are significantly different. Third, the *F*-test evaluates whether the variance explained by the dependent variable is greater than the unexplained variance caused by the grouping variable. A non-significant *F*-test suggests the model is too weak to be reliable, meaning there is more noise than explained variance. If all groups are similar, the *F*-value will be small, but if they differ significantly, the *F*-value will be larger. The *F*-test is often called the *F* ratio because it is calculated as a ratio of the variance due to group differences to the variance due to individual differences. This ratio helps determine whether further analysis is needed to pinpoint where the group differences occur.

Test Your Understanding 12.2 – Does the *F*-test tell us specifically between which groups differences exist within data?

There is quite a bit of information that goes into calculating the *F*-test. Try not to get bogged down in the next section, as we outline the different components that make up the test. The basic formula for the *F*-test is

$$F = \frac{\text{between groups variance}}{\text{within groups variance}}$$

or the more "statistical" formula is $F = \frac{MS_M}{MS_R}$

Let's break down the equation(s) for the *F*-test, as it's critical to know what the ANOVA is doing. The mean squares (MS) are calculated by dividing the respective sums of squares by their associated degrees of freedom, and they serve as the basis for the *F* ratio. There are four key terms that we need to know to be able to calculate the *F*-test: total sum of squares (SS_T), model sum of squares (SS_M) (sometimes called the between groups sum of squares, SS_B), residual sum of squares (SS_R) (sometimes called within group sum of squares, SS_W), and mean squares (MS_M). The **total sum of squares (SS_T)** is the total variation within our entire data. To calculate the total variation within your data, compute the difference between the grand mean (total of all data values/sample size) and each (individual) data point, square the difference, and then add it all up. The equation follows below, but don't worry, it's computed by statistical programs for us. We will explain all the symbols as well:

$$SS_T = \sum_{i=1}^{N} \left(x_i - \bar{x}_{\text{grand}}\right)^2$$

The following is what each statistical symbol means above:

SS_T = Total sum of squares
N = sample
Σ = Greek letter sigma, which means to sum up or add up
x_i = each observed data point
$\bar{x}_{\text{grand}}$ = grand mean

Let's say we are trying to understand the extent to which a new migraine medication is effective. So, the dependent variable is the effectiveness of the medication (pain relief). There are three groups (independent variable): Group One (receives the medication), Group Two (receives aspirin), and Group Three (receives a sugar pill). The SS_T would tell us the total amount of variation within the entire sample on pain relief.

Test Your Understanding 12.3 – Is the total sum of squares the total variation within our sample or within a particular group?

Our next step is to understand the **model sum of squares (SS_M)**. The SS_M tells us how much variation in our dependent variable (pain relief) is explained by scores coming from different groups. The equation is below, and is like the SS_T equation, except the values are not individual data points, but group means:

$$SS_M = \sum_{g=1 n_g}^{k} \left(\bar{x}_g - \bar{x}_{\text{grand}}\right)^2$$

The following is what each statistical symbol means above:

SS_M = Model sum of squares
k = Number of groups
g = group
n_g = size of group
$\bar{x}_g$ = group mean

Based on this equation, we could calculate the variation coming from each individual group: aspirin, sugar pill, and the medication.

Now that we know the total variation in the sample and how much each group explains, the **residual sum of squares (SS_R)** explains how much of the variation is not explained by our model. Whenever we conduct a statistical test, there will always be unexplained things, due to statistical error, and because we did not measure all possible variables that might influence our dependent variable. Whenever you conduct an experiment, you will (we hope) do your due diligence and try to measure all the necessary variables. However, there will be some things that we simply cannot predict or know, and we also can't always predict how people will behave during our experiments. To calculate the $SS_{R,}$ we compare the score obtained by a person (or thing) and the mean of the group from which that person (thing) belongs. The formula looks like the previous two you have seen, but you will see an individual and a group being compared:

$$SS_R = \sum_{g=1n_g}^{k} \left(\bar{x}_g - \bar{x}_{\text{grand}}\right)^2$$

where X_{ig} is each score in a group.

Calculating the SS_R tells us how much variation is unexplained when we try to compute differences in migraine pain relief between our three groups. Finally, we can compute the mean square (MS_M), or the total variation that our experimental manipulation explains. This might seem a bit confusing at first, particularly since the SS_T also explained total variation, but let us explain. The MS_M is an average sum of squares. To generalize to a population, we divide the SS_M by our degrees of freedom (*df*), and the MS_M is the average amount of variation that our model explains. Thus, the formula for MS_M is:

$$MS_M = \frac{SS_M}{df_M}$$

In addition, we can divide the SS_R by the *df* to also get our average amount of variation that is unexplained:

$$SS_R = \frac{SS_R}{df_R}$$

When we put it all together, we get the *F*-test equation from a few pages earlier, which essentially explains the amount of variation that our model explains / the amount that is unexplained or unaccounted for:

$$F = \frac{MS_M}{MS_R}$$

There are a few final things to consider when interpreting the *F*-test. First, as the *F*-test is an omnibus test, it only tells us there is a significant difference somewhere among our groups. It does not indicate which groups the difference is between. Thus, follow-up post-hoc tests are often needed to fully interpret the data. Second, each of your authors has had students ask them why they can't run multiple *t*-tests instead of trying to interpret this *F*-test. When you have multiple groups, and you run comparisons such as (G_1 vs. G_2), (G_1 vs. G_3), and (G_2 vs. G_3), you increase your risk of committing a Type I error – think back to Chapter 10. The *F*-test is an omnibus test that tests all the parameters at once, thus reducing your risk of a Type I error. (In other words, the *F*-test ensures a single overall confidence level for all group comparisons, whereas multiple *t*-tests only provide confidence for individual comparisons, increasing the risk of Type I error.) Third, like with other tests you have studied, homogeneity of variance is assumed. However, if homogeneity of variance is "violated," there are some corrections we can run that alter the *F* equation. Welch's *F* and the Brown–Forsythe *F* are commonly used corrections. Both tests are relatively effective at managing Type I errors, but Welch's *F* has more statistical power.

Test Your Understanding 13.4 – Why are you likely to commit a Type 1 error if you run a series of *t*-tests instead of an ANOVA when comparing multiple groups?

Post-Hoc Procedures

So, let's return to the Jedi Council. All the Jedi decided to use their midi-chlorian count to determine who is the strongest Jedi.[1] All the members just had their midi-chlorians checked to see who is the most Force-sensitive among them. The result comes back that there is a significant difference among the Council, the *F*-test. While the significant *F*-test tells them there are differences among the Jedi Council members, it doesn't tell them where (with whom) the differences are. **Post-hoc comparisons** (sometimes called pairwise or multiple comparisons) are tests that identify specific mean differences between groups in an ANOVA. These tests are typically only used when you have a significant *F* result. You can use them even when your *F* is not significant; however, the purpose of these tests is to find out more information about the significant differences between the groups in your ANOVA.

There are multiple types of post-hoc comparisons. When choosing the most appropriate comparison, there are a few things to consider. First, does the test reduce your risk of a Type I error? If you remember from Chapter 7, a Type I error is when you incorrectly reject the null hypothesis. When choosing a post-hoc comparison, the extent to which the test is "liberal" or "conservative" can influence your chance of committing a Type I error. A more "conservative" test will have a lower minimum *p*-value threshold (maybe $p < .01$), while a more "liberal" test might be more generous for significance ($p < .05$). The conservative test, with its tighter control of significance, reduces the chances of committing a Type I error. Notably, all post-hoc tests are less likely to result in Type 1 error than a series of *t*-tests because they calculate significance holistically rather than one test at a time.

Second, does the test reduce your risk of a Type II error? This is when you fail to reject the null hypothesis when it is, in fact, false. Tests with greater power are better equipped to detect differences.

Third, how reliable is the test when homogeneity of variance is violated? Questions such as how the test operates when the sample sizes are different between the groups, when the data is not normally distributed, and when the variances are unequal (very) all impact homogeneity. In these cases, you do not have to worry too much, as most post-hocs will perform well with minor statistical deviations on these questions. However, when there are large deviations within and between the groups, you will need to use specific tests.

The first question to ask yourself when deciding on which post-hoc to use is whether you have a lot of variance within and between your groups. If you don't (think back to Chapters 4–5, where you learned how to visualize and check variance), then choose one of the tests in Table 12.1. If you do, then use one of the tests in Table 12.2. These tests are not inclusive of all the post-hoc comparisons you will encounter in SPSS. However, these are the most used comparisons.

You will notice there are a *lot* of options above for when equal variances are assumed. The authors of this book have normally used Tukey, Scheffé, or Bonferroni. When we have had equal sample sizes, we have tended to use Scheffé because it is more conservative. However, as we often do not have 100% equal sample sizes, but still have similar or close to similar variances, we tend to use Tukey most often.

Of these four tests, the most used one among the authors is the Games–Howell. It is the most powerful.

Table 12.1 Equal Variances Assumed Post-Hoc Comparison Tests

Test	*Description*	*Usage*
Bonferroni	Most conservative test and easy to use. Not a very powerful test because it over-corrects for Type I errors.	When you want to be guaranteed that you do not commit a Type I error. Weak at detecting a difference if one does exit.
Dunn	Conservative test and lacks power, particularly when you have many groups.	Use when comparing a small number of groups.
Dunnett	Designed to compare a control group to treatment groups.	Use this to compare a control group to treatment groups.
Gabriel's	Designed for use when group sizes are unequal. A powerful liberal test.	Use this test when sample sizes are unequal.
REGWQ (Ryan, Einot, Gabriel, and Welsch Q)	Powerful in controlling Type I error. Do not use when group sizes are different.	Use when you have equal variances and group sizes.
Scheffé	Best for general comparisons. More conservative than Tukey but lacks power.	Use instead of the Tukey test when the group sizes are unequal.
Tukey	A conservative test. More powerful than Bonferroni and has more power than Scheffé and Dunn.	Use when you have equal variances and sample sizes.

Table 12.2 Equal Variances Not Assumed Post-Hoc Comparison Tests

Test	*Description*	*Usage*
Games–Howell	Most powerful of these tests. Can be liberal, though, when sample sizes are small. Effective when group sizes are unequal	Use when homogeneity of variances is violated
Dunnett's C	Slightly more conservative than Games–Howell	Use when homogeneity of variances is violated
Dunnett's T2	Slightly more conservative than Games–Howell	Use when homogeneity of variances is violated

ANOVA With Multiple Independent Variables

Up to this point, we have been talking about one-way ANOVA. A one-way ANOVA has only one independent variable. You can also run ANOVAs with multiple dependent variables. These types of ANOVAs are what you run when you want to see if some combination of independent variables affects the dependent variable. For example, Kelly et al. (2024) conducted an experiment to see how the use of various Zoom backgrounds affected students' perception of instructor credibility. Because prior research had found that male and female students responded to instructors' use of technology differently, they wanted to know not just if perceptions of instructor credibility were found, but whether those differences were specific to student gender. To do this, they had to run a **factorial ANOVA**. Their factorial ANOVA was 2 x 3, meaning there were two groups of genders analyzed, and three groups of backgrounds analyzed. So, their data broke down as shown in Table 12.3:

Table 12.3 Experiment Groups

	Beach Background	*Architecture Background*	*University Background*
Male	Males who saw the Beach Background	Males who saw the Architecture Background	Males who saw the University Background
Female	Females who saw the Beach Background	Females who saw the Architecture Background	Females who saw the University Background

Their results showed both a main effect and an interaction effect. A **main effect** means that two of the factors were different, where a factor is one of the variable categories (male vs. female OR beach vs. architecture vs. university). The main effect they found was that across data, females considered the instructors to be more credible, regardless of their background. There was no main effect for background. There was, however, an **interaction effect**, an effect in which the combination of categories mattered. The interaction effect was that male students rated the credibility especially low in the beach and architecture backgrounds, meaning that males only found the instructor credible when using a university background on Zoom. This was an important finding because it meant that, for the sake of male students' perceptions, instructors need to use a university-branded background.

How to Run an ANOVA in Excel and SPSS

The data for this chapter are 80 citizens that the Jedi Council surveyed across five planets: Coruscant ($n = 31$), Tatooine ($n = 11$), Naboo ($n = 14$), Jakku ($n = 10$), and Kashyyk ($n = 14$). Each citizen had their Jedi abilities measured on a scale of 0–100.

ANOVA in Excel

As with the last chapter, we need to emphasize that Excel is not the right tool for this analysis. It does not check assumptions for you, nor provide options for post-hoc analyses. However, if you want to conduct a simple comparison of groups to see if you have a significant F-statistic in Excel, here are the steps:

1. Organize your data so that your grouping variable is designated with column or row headers
2. Call the Data Analysis tool as you have for prior analyses
3. Select Single Factor ANOVA
4. Click Okay
5. Type the range of cells containing your data in Range (It's okay if this includes blank cells because you have more data in some rows or columns)
6. Select Rows or Columns to tell Excel which way your data is organized
7. If your data has row or column headers, tell Excel this by clicking "Labels"
8. Provide Excel with the first blank cell you wish the output to be provided in
9. Click Okay

ANOVA in SPSS

Running an ANOVA in SPSS is rather simple. First, as shown in Figure 12.1, open the One-way ANOVA menu by clicking:

1. Analyze
2. Compare Means and Proportions
3. One-Way ANOVA
4. Click on your Dependent variable (what you are measuring) and move it over to the dependent list category. In this case, we are measuring force ability among citizens of different *Star Wars* planets
5. Click on the factor or group in which you will compare force ability and move it to Factor. (See Figure 12.2.) In this case, it is "planet"
6. You now need to determine which type of post-hoc you want to run. Think back to earlier in the chapter where we discussed how the ANOVA will report an F, showing whether differences exist between the groups, but not where the differences are. Before you choose to run a post-hoc, you should look at the data and understand it. In this case, the data includes participants from five planets: Coruscant ($n = 31$), Tatooine ($n = 11$), Naboo ($n = 14$), Jakku ($n = 10$), and Kashyyk ($n = 14$). These are relatively small samples, and if you run descriptive analyses on this data (think back to Chapter 5–6), equal variances are not assumed. Thus, a Games–Howell post-hoc is the most suitable post-hoc test to run. Thus, in the One-Way ANOVA box, click on Post-Hoc (noted in Figure 12.2), and you will open up the post-hoc option box, shown in Figure 12.3. Choose Games–Howell and then press Continue

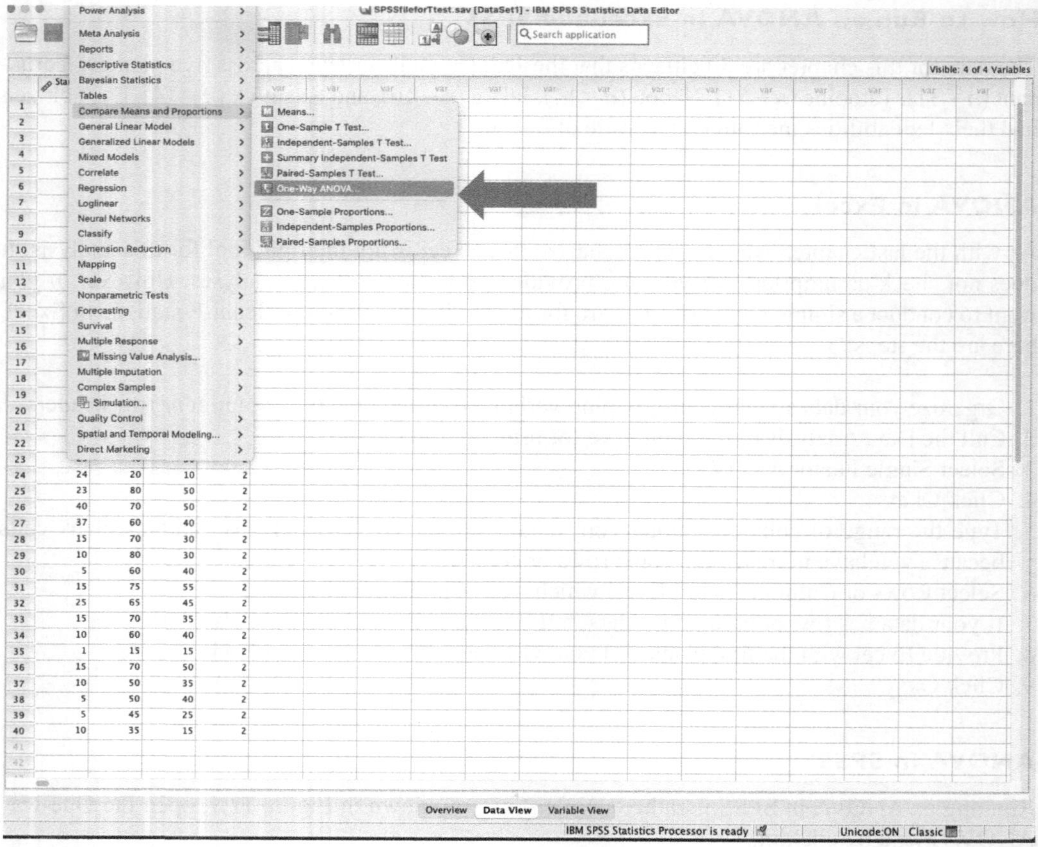

Figure 12.1 Calling up the ANOVA in SPSS.

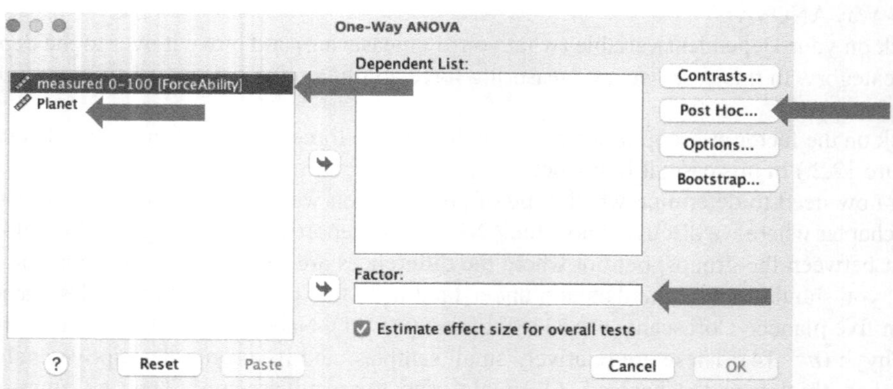

Figure 12.2 One-Way ANOVA box.

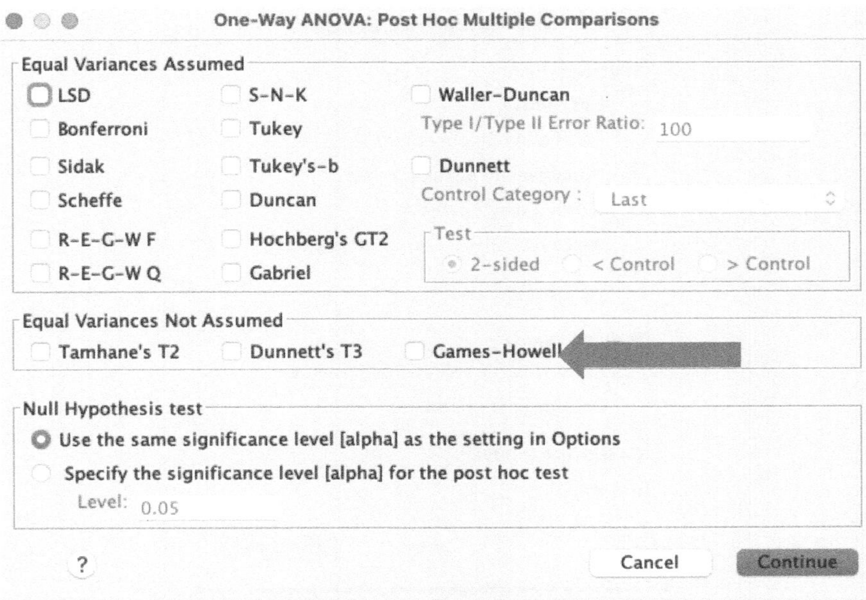

Figure 12.3 Post-Hoc box in SPSS.

7. Next, choose the Options from the main ANOVA box and select Descriptive, Fixed and random effects, Homogeneity of variance test, Brown–Forsythe test, Welch test, and Means plot. (See Figure 12.4.) Then click Continue

You will get the following outputs, which are important for you to have when writing up your results. Table 12.4 is the Descriptive Statistics Output. This output shows the means, standard deviations, sample size, and standard errors for each sample (planet).

The next output, Table 12.4 is the Test of Homogeneity of Variances. As discussed earlier, when we run an ANOVA, it is critical all samples should have equal or similar levels of variance. However, if this is not the case, there are corrections that can be conducted and different post-hocs we can run. As shown in Table 12.5, Levene's statistic is significant, showing there is significant variation between the samples; thus, homogeneity of variance cannot be assumed. Therefore, running Games–Howell was an appropriate post-hoc.

The next output is the ANOVA result (Table 12.6). In this output you have the *F*-statistic, the significance of the overall ANOVA, the *df*, and all the data that makes up the *F* (between and within groups variance). The key to this output is that it tells you if the ANOVA is significant, if there is significant difference within your sample. In this case there is, $F = 3.82$, $p = .007$.

The next output reports the effect sizes present in the ANOVA. An effect size if one way to quantify difference between groups. In paired samples *t*-tests we reported the correlation (*r*). In ANOVA there are different estimations of effects that you can report. The authors of this text typically report **eta-squared** (η^2). Eta-squared measures the proportion of variance linked with each main effect in an ANOVA. Its value ranges from 0 to 1. Higher values mean

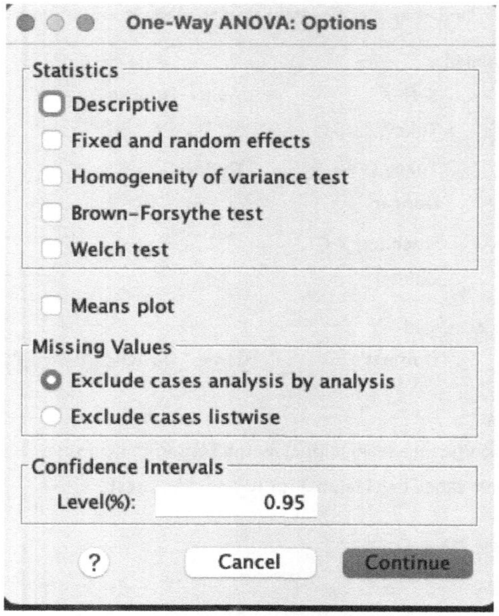

Figure 12.4 Post-Hoc box in SPSS.

Table 12.4 ANOVA Descriptives

measured 0–100										
		N	*Mean*	*Std. Deviation*	*Std. Error*	*95% Confidence Interval for Mean*		*Minimum*	*Maximum*	*Between-Component Variance*
						Lower Bound	*Upper Bound*			
Coruscant		31	32.84	22.582	4.056	24.56	41.12	1	80	
Tatooine		11	17.36	27.721	8.358	−1.26	35.99	1	100	
Naboo		14	36.43	26.849	7.176	20.93	51.93	10	95	
Jakku		10	14.50	8.822	2.790	8.19	20.81	2	30	
Kashyyk		14	14.64	8.196	2.190	9.91	19.37	5	30	
Total		80	25.86	22.739	2.542	20.80	30.92	1	100	
Model	Fixed Effects			21.269	2.378	21.13	30.60			
	Random Effects				5.147	11.57	40.15			84.710

Table 12.5 Tests of Homogeneity of Variances

		Levene Statistic	*df1*	*df2*	*Sig.*
measured 0–100	Based on Mean	3.610	4	75	.010
	Based on Median	1.959	4	75	.109
	Based on Median and with adjusted df	1.959	4	47.195	.116
	Based on trimmed mean	3.185	4	75	.018

Table 12.6 ANOVA Output

measured 0–100

	Sum of Squares	*df*	*Mean Square*	*F*	*Sig.*
Between Groups	6919.606	4	1729.901	3.824	.007
Within Groups	33927.882	75	452.372		
Total	40847.487	79			

Table 12.7 ANOVA Effects Sizes[a,b]

		Point Estimate	*95% Confidence Interval*	
			Lower	*Upper*
measured 0–100	Eta-squared	.169	.016	.285
	Epsilon-squared	.125	−.037	.247
	Omega-squared Fixed-effect	.124	−.036	.244
	Omega-squared Random-effect	.034	−.009	.075

a. Eta-squared and Epsilon-squared are estimated based on the fixed-effect model.
b. Negative but less biased estimates are retained, not rounded to zero.

more of an effect, or variation. Researchers typically rely on the following levels to interpret eta-squared:

.01 to.06 = small effect
.06 to.14 = medium effect size
.14 or higher = large effect size

Essentially, an eta-squared of .24 means there is a vast amount of variation in the dependent variable related to one of the groups in the ANOVA. According to Table12.7, the ANOVA Effects Sizes output, there is a large effect on the dependent variable, $\eta^2 = .17$.

The next output is the Robust Tests of Equality of Means (Table 12.8). This output includes the result of the Welch and Brown–Forsythe tests. Both tests compare the means of populations and report if there are significant differences. In this case, both tests report there are significant mean differences between some groups, but the tests do not say between which groups.

Table 12.8 Robust Tests of Equality of Means

measured 0–100				
	Statistic[a]	*Df1*	*df2*	*Sig.*
Welch	5.602	4	30.353	.002
Brown–Forsythe	4.086	4	38.162	.007

a. Asymptotically F distributed.

Table 12.9 Multiple Comparisons

Dependent Variable: measured 0–100

Games–Howell

(I) Planet	*(J) Planet*	*Mean Difference (I-J)*	*Std. Error*	*Sig.*	*95% Confidence Interval*	
					Lower Bound	*Upper Bound*
Coruscant	Tatooine	15.475	9.290	.482	−13.22	44.17
	Naboo	−3.590	8.243	.992	−28.08	20.90
	Jakku	18.339*	4.923	.005	4.23	32.45
	Kashyyk	18.196*	4.610	.003	5.06	31.33
Tatooine	Coruscant	−15.475	9.290	.482	−44.17	13.22
	Naboo	−19.065	11.016	.437	−51.84	13.71
	Jakku	2.864	8.812	.997	−25.15	30.88
	Kashyyk	2.721	8.640	.998	−25.06	30.50
Naboo	Coruscant	3.590	8.243	.992	−20.90	28.08
	Tatooine	19.065	11.016	.437	−13.71	51.84
	Jakku	21.929	7.699	.073	−1.55	45.40
	Kashyyk	21.786	7.503	.069	−1.31	44.88
Jakku	Coruscant	−18.339*	4.923	.005	−32.45	−4.23
	Tatooine	-2.864	8.812	.997	−30.88	25.15
	Naboo	−21.929	7.699	.073	−45.40	1.55
	Kashyyk	−.143	3.547	1.000	−10.83	10.55
Kashyyk	Coruscant	−18.196*	4.610	.003	−31.33	-5.06
	Tatooine	−2.721	8.640	.998	−30.50	25.06
	Naboo	−21.786	7.503	.069	−44.88	1.31
	Jakku	.143	3.547	1.000	−10.55	10.83

*. The mean difference is significant at the 0.05 level.

The next output is the post-hoc comparisons or multiple comparisons table (Table 12.9). This output shows where differences are between the groups. As we did not have equal variances in this data, and the dataset was rather small, we ran a Games–Howell post-hoc. Based on this post-hoc procedure. We can see differences between the groups, identified by as asterisk (*). For example, individuals from Coruscant are significantly higher in force abilities than individuals from Jakku and Kashyyk. The other groups do not differ significantly from one another.

The final output is a means plot, which visually depicts the means for you. This is not something that is usually reported or provided in a journal article or a submission. However, it is helpful to visualize your data.

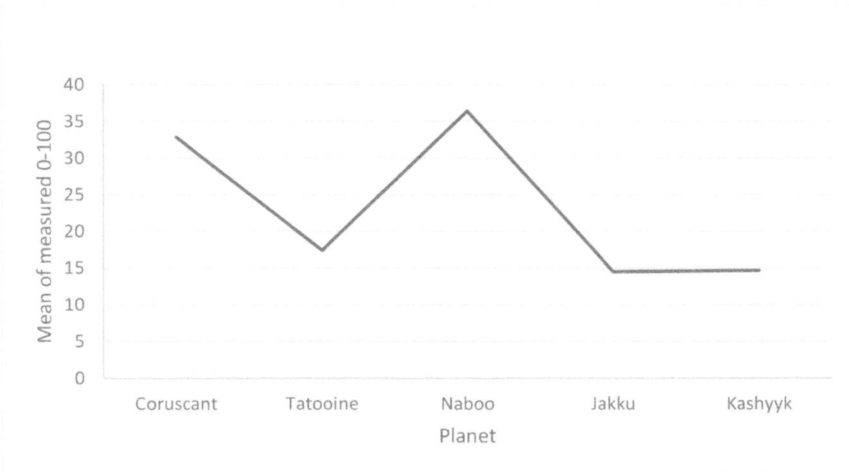

Figure 12.5 Means plots.

In writing up this ANOVA, it could be done as follows. The Jedi Council set forth to explore the extent to which citizens on different planets differed in force abilities. Citizens were tested on five planets. A one-way ANOVA was calculated using planet as the independent variable and force ability as the dependent variable. A significant difference was noted: $F(4, 75) = 3.82$, $p = .007$, $\eta^2 = .17$. In a follow-up, a Games–Howell post-hoc was conducted. The Games–Howell post-hoc indicated citizens of Coruscant ($M = 32.84$, $SD = 22.58$) were significantly higher in force abilities than citizens from Jakku ($M = 14.50$, $SD = 8.82$) and Kashyyk ($M = 14.64$, $SD = 8.20$).

Factorial ANOVA in SPSS

To run a factorial ANOVA in SPSS, the steps are going to be very similar to what you've already done. The main difference is where you go in the menu. Start, as shown in Figure 12.6, by going to:

1. Analyze
2. General Linear Model
3. Univariate

From there, you will repeat the steps you've learned in one-way ANOVA. Select a post-hoc analysis, choose the options for tests of homogeneity and descriptive statistics. It is also recommended that you choose the option to produce Estimates of Effect size, which you will learn more about in Chapter 13.

Summary

In practice, researchers use ANOVA to compare the means across more than two groups. The math in an ANOVA does not actually compare means, though – it compares variances between groups. The *F*-test indicates if there is a difference in how and where data are spread across two or more samples, considering where it clusters and how much spread those clusters are.

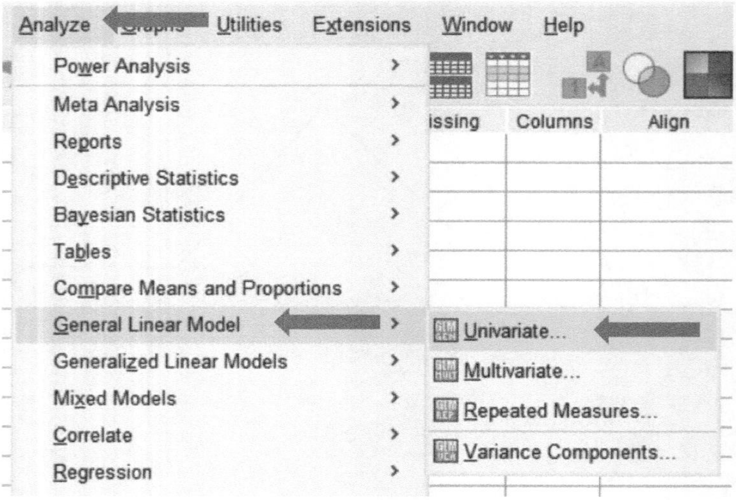

Figure 12.6 Univariate.

Note

1 For those *Star Wars* fans out there, we do not need to explain what midi-chlorians are. However, in the *Star Wars* universe, midi-chlorians are microscopic life forms that live in all living organisms. These life forms have a symbiotic relationship with the Force and the life form. Essentially, those with a higher midi-chlorian count are more in touch with or more powerful in the Force. Notably, though, as you learned in your math class, you do not need to know WHAT x, your variable, is in order to do calculations about it.

References

Croucher, S. M., Yotes, T., Ashwell, D., & Condon, S. (2024). A comparison of the relationship between extraversion and argumentativeness in the United States, Canada, Australia, and New Zealand. *Communication Research Reports*, *41*(2), 82–93. https://doi.org/10.1080/08824096.2024.2327548

Kelly, S., Kim, J., Goke, R., & Bryant, K. (2024). Do instructors' choice of virtual background in online classes matter? Effects of personal vs. professional virtual background use on instructor credibility and student learning. *International Journal of Human–Computer Interaction*, *40*(22), 7314–7321. https://doi.org/10.1080/10447318.2023.2263695

Chapter 13

Multiple Analysis of Variance (MANOVA)

What Will I Learn in This Chapter?

It is no secret that a boss like Darth Vader is hard to please. If it's not done before he even asks, it's *too slow.* Let's say the Empire is building another Death Star, and Vader wants to know which team is the most cost-effective: Engineers, Stormtroopers, or Sith Apprentices. From what we've learned from the previous chapter on ANOVA, you thought you could satisfy Vader by proposing that running an ANOVA comparing all three groups simultaneously is more efficient and robust than performing multiple *t*-tests. While you were awaiting a compliment from him, Vader emotionlessly responded, "I find your statistical approach . . . acceptable. But I also require an analysis of their faith in this project, for a lack of faith . . . would be most disturbing." You are now panicking because this scenario goes beyond one-way ANOVA, as there are multiple dependent variables now. Should you propose performing multiple one-way ANOVAs? That takes too long and may

DOI: 10.4324/9781003500117-13

incur other problems! What you will learn in this chapter might save you in the said situation, as we will cover:

- What is a MANOVA
- Assumptions of MANOVA
- How to run a MANOVA in SPSS
- How to interpret and report MANOVA results

What Is MANOVA?

In the previous chapter, we've learned that ANOVA is a powerful tool to test for differences in a single dependent variable across groups. In research projects, there is often a need to look at multiple dependent variables at the same time. Should we run multiple ANOVAs in this case? You might have guessed that the answer is no, and the rationale is like why we shouldn't run multiple *t*-tests as a replacement for ANOVA, as that would inflate our chance of making Type I errors. In a situation where group differences are being tested on several dependent variables simultaneously, an extension of ANOVA that is called **MANOVA** should be used. Adding M in front of ANOVA denotes that this is a **multivariate analysis of variance**.

In addition to being simply more efficient than ANOVA, MANOVA essentially addresses the question: When considered together, are the differences in dependent variables significant across groups? This is important because the dependent variables are often interrelated in real-world problems and their intercorrelations may skew the results. It is entirely possible for individual ANOVA results to show significant differences for some dependent variables while the combined MANOVA result is insignificant. Like ANOVA versus a series of *t*-tests or regression versus a series of correlations, MANOVA provides confidence for all predictions collectively rather than for individual predictions. For example, in a MANOVA with two dependent variables, a 95% confidence level means there is 95% confidence that the model's predictions about both variables, considered together, are accurate – not 95% confidence for each variable independently.

An example of communication research that has used MANOVA is Mikkelson et al.'s (2011) study that looked into siblings' use of relational maintenance behaviors. The authors aimed to compare genetically related siblings to nongenetically related siblings on their relational maintenance behaviors. Since relational maintenance behaviors are comprised of multiple behaviors such as positivity, assurances, openness, networks, and tasks, MANOVA is an appropriate choice to include these dependent variables in one analysis.

<u>**Test your understanding 13.1**</u>: What are the benefits of performing a MANOVA instead of multiple ANOVAs?

MANOVA Assumptions

Like other tests, MANOVA makes certain assumptions that must be met to ensure the validity of the results. MANOVA has similar assumptions to ANOVA that are extended to reflect the multivariate nature of the analysis. Here are the key assumptions:

1. *Multivariate normality*: In ANOVA, the dependent variable is assumed to be normally distributed within each group. Here, each dependent variable and all linear combinations of dependent variables must follow a multivariate normal distribution within groups.

2. *Homogeneity of covariance matrices:* In addition to this assumption, MANOVA also requires the homogeneity of covariance matrices across groups, meaning that the relationships between dependent variables remain consistent across all levels of the independent variable. Box's M test of equality of covariance matrices is used to test this assumption in MANOVA.
3. *Independence of observations*: Observations should be independent of each other.
4. *Random sampling*: Data should be randomly sampled from the population of interest and measured at an interval level.
5. *Sample size*: MANOVA needs larger sample sizes, especially when the number of dependent variables increases, to provide stable estimates of covariance. A rule of thumb is to aim for 20–30 observations per group per dependent variable. For instance, a comparison between two groups on three dependent variables would require at least 20 × 2 × 3 = 120 participants.

Next, let's try to run a MANOVA with an example. Please note that the simplified data below is merely used as a demonstration of how to perform MANOVA in SPSS. Aiming for at least 20 observations per group per dependent variable is still a desirable goal when running MANOVA.

Say your employer wants to explore how *team structure* impacts *employee performance*, which has two subdimensions: *engagement* and *productivity*. In the organization, there are three team structures comprised of employees with different percentages of working onsite:

- *Virtual Team* (entirely remote),
- *Face-to-Face Team* (entirely in person),
- *Hybrid Team* (combination of virtual and face-to-face).

After six months in their respective team structures, employees are surveyed on their levels of engagement and productivity as shown in Table 13.1:

In this example, we have two dependent variables that are employee engagement and productivity, and they are likely intercorrelated, as we know employees tend to be more productive with higher engagement. So, MANOVA would be the perfect choice here.

Table 13.1 The Influence of Team Structure on Employee Performance

Employee Engagement			*Employee Productivity*		
Virtual	*F2F*	*Hybrid*	*Virtual*	*F2F*	*Hybrid*
7	6	8	8	7	9
8	8	8	7	8	8
7	7	9	7	7	7
7	5	7	5	6	7
6	8	9	8	6	8
8	7	7	7	8	5
7	8	8	6	5	9
6	8	8	8	8	8
6	9	9	7	6	9
5	7	7	6	8	8

Performing MANOVA in SPSS

Since Excel does not have a built-in function for MANOVA, it would require some manual computation and/or additional programming, which can be a more time-consuming and error-prone process. Additionally, Excel does not provide tools for diagnostic tests, and its output is much less comprehensive. So, we will only cover how to perform MANOVA in SPSS in this chapter.

1. First, we can enter the data into SPSS, as shown in Figure 13.1. For team structure, we've coded *Virtual* as 1, *F2F* as 2, and *Hybrid* as 3. MANOVA can be found by going to Analyze General Linear Model Multivariate
2. After clicking on Multivariate, you will be taken to the interface shown in Figure 13.2, where we can move Engagement and Productivity to Dependent Variables and Team Structure to Fixed Factor(s)
3. Now, we have a few things to specify before hitting the OK button. Notice there is a column of buttons on the right side of the table. First, we can go to Options and select Descriptive statistics, Estimates of effect size, and Homogeneity tests, as these will provide vital information on the overall model (See Figure 13.3)

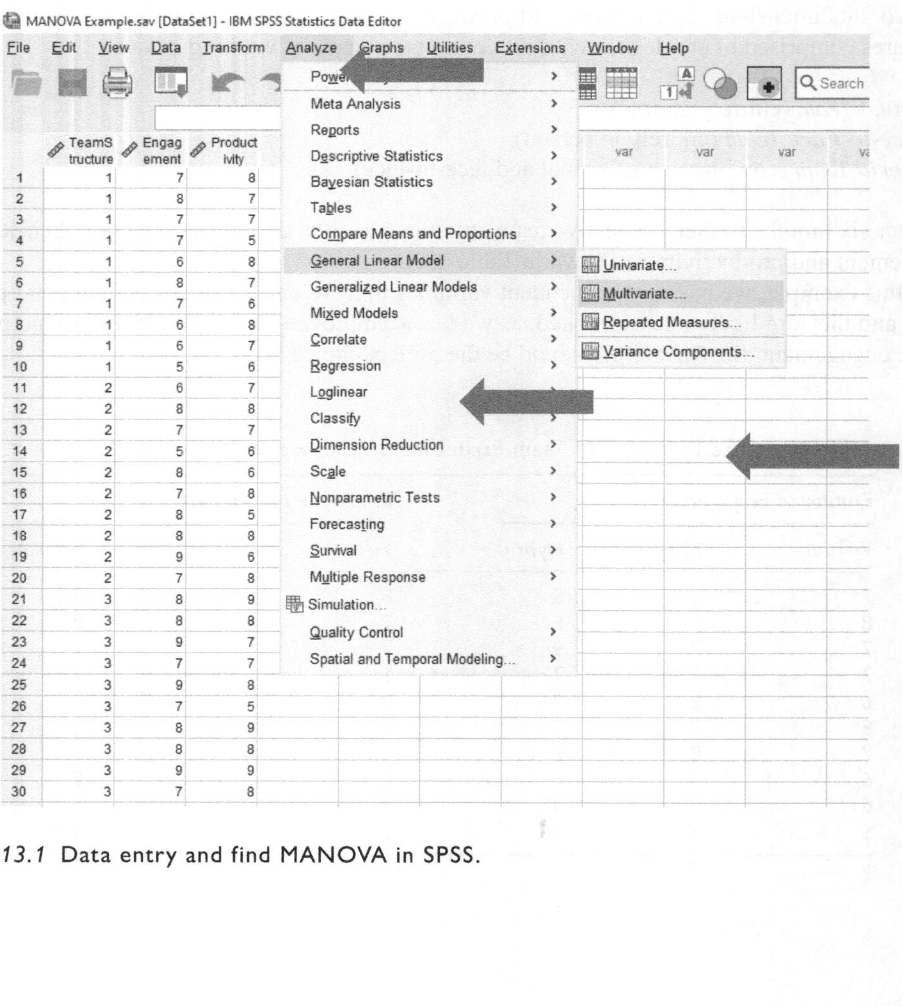

Figure 13.1 Data entry and find MANOVA in SPSS.

Figure 13.2 MANOVA specification.

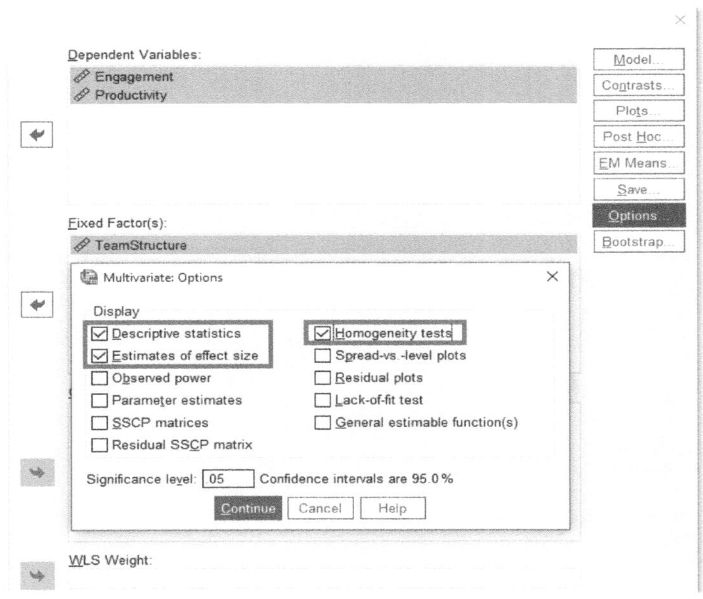

Figure 13.3 Options.

4. Click Continue
5. While we are on this page, we also want to specify Contrasts and Post-Hoc, the two important follow-up analysis methods. Select Contrasts and change the contrast method to Simple (Figure 13.4)
6. Click Continue
7. Select Post-Hoc . . . The Post-Hoc interface is very similar to what we've seen in ANOVA, and you can choose your preferred analysis. You can always refer back to that chapter to determine which post-hoc test should be chosen for a given dataset. We will only select Tukey here to help simplify things

Testing the Homogeneity Assumption

After selecting Homogeneity tests, SPSS will perform two tests as shown in Table 13.2. Levene's test should be familiar as that is what we use in ANOVA to test for equality of variances for the dependent variable. As we have two dependent variables in this MANOVA, SPSS has simply conducted two Levene's tests on both engagement and productivity. The non-significant values show that the variance of both dependent variables is equal across all groups. That is a great start!

Since we are performing a MANOVA, the assumption of equality of covariance matrices needs to be met. Box's M test has been used to test whether the covariance matrices of the dependent variables are equal across the groups. Like Levene's test in ANOVA, a non-significant result indicates that the covariance matrices are equal across groups. In this example, Box's M value is non-significant (p =.82), meaning the assumption is met.

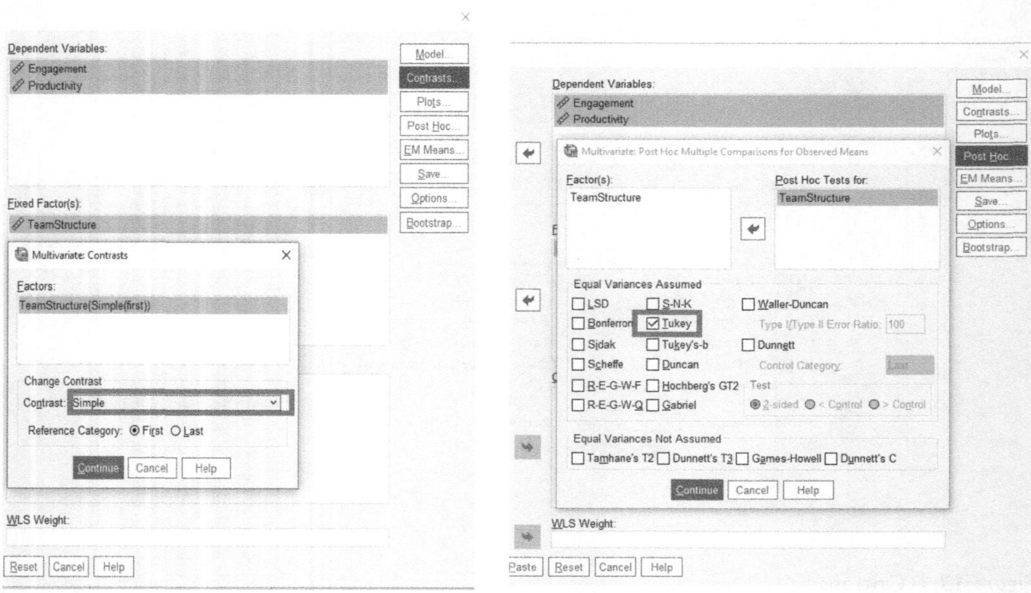

Figure 13.4 Contrasts and Post-Hoc.

Table 13.2 Tests of Homogeneity

Box's Test of Equality of Covariance Matrices[a]

Box's M	3.256
F	.484
df1	6
df2	18168.923
Sig.	.820

Tests the null hypothesis that the observed covariance matrices of the dependent variables are equal across groups.
a. Design: Intercept + TeamStructure

Levene's Test of Equality of Error Variances[a]

		Levene Statistic	*df1*	*df2*	*Sig.*
Engagement	Based on Mean	.697	2	27	.507
	Based on Median	.578	2	27	.568
	Based on Median and with adjusted df	.578	2	25.502	.568
	Based on trimmed mean	.690	2	27	.510
Productivity	Based on Mean	.206	2	27	.815
	Based on Median	.185	2	27	.832
	Based on Median and with adjusted df	.185	2	23.119	.832
	Based on trimmed mean	.193	2	27	.826

Tests the null hypothesis that the error variance of the dependent variable is equal across groups.
a. Design: Intercept + TeamStructure

Interpreting MANOVA Output

SPSS produces four lines of results when performing a MANOVA analysis, which can be intimidating and confusing. These are simply different test statistics that differ in test power and robustness, and we often only need to choose one to report in our research. Another good news is that the differences among these test statistics are relatively small, especially when the sample size is small to medium (Stevens, 1980). It is also evident in the following table that all four test statistics showed significant results despite some small differences. To help simplify this decision, we will only talk about the two most popular choices: Wilks' lambda (λ) and Pillai's Trace (V), as the other two are mostly used in special cases. When the sample size is large and all the assumptions are met, Wilks' lambda is often the go-to choice because it is sensitive to sample size, and its performance improves as the sample size gets larger. When there is concern about assumption violations, unequal group sizes, or heterogeneity of covariances, researchers may opt for Pillai's Trace for its robustness.

To mirror other communication research, we choose to report Wilks' lambda (this result is in Table 13.3) in this example: Wilks' λ =.69, F= 2.71, p =.04. Generally, a smaller Wilks' lambda value suggests a greater difference between groups, whereas there is no difference between groups when Wilks' lambda value equals 1. Here, our Wilks' lambda value is .69 and also significant, which indicates that at least one of the groups differs significantly from the others in terms of the combination of dependent variables. In other words, the employee engagement and productivity levels across the three team structures are not the same. While it is exciting to find out that some

Table 13.3 Multivariate Test Results

Effect		*Value*	*F*	*Hypothesis df*	*Error df*	*Sig.*	*Partial Eta-Squared*
Intercept	Pillai's Trace	.990	1277.779[b]	2.000	26.000	<.001	.990
	Wilks' lambda	.010	1277.779[b]	2.000	26.000	<.001	.990
	Hotelling's Trace	98.291	1277.779[b]	2.000	26.000	<.001	.990
	Roy's Largest Root	98.291	1277.779[b]	2.000	26.000	<.001	.990
TeamStructure	Pillai's Trace	.323	2.596	4.000	54.000	.046	.161
	Wilks' lambda	.685	2.705[b]	4.000	52.000	.040	.172
	Hotelling's Trace	.448	2.802	4.000	50.000	.036	.183
	Roy's Largest Root	.422	5.693[c]	2.000	27.000	.009	.297

a. Design: Intercept + TeamStructure
b. Exact statistic
c. The statistic is an upper bound on *F* that yields a lower bound on the significance level.

Table 13.4 Univariable ANOVA Test Results

Tests of Between-Subjects Effects

Source	*Dependent Variable*	*Type III Sum of Squares*	*df*	*Mean Square*	*F*	*Sig.*	*Partial Eta-Squared*
Corrected Model	Engagement	8.467[a]	2	4.233	4.363	.023	.244
	Productivity	5.400[b]	2	2.700	2.183	.132	.139
Intercept	Engagement	1613.333	1	1613.333	1662.595	<.001	.984
	Productivity	1555.200	1	1555.200	1257.198	<.001	.979
TeamStructure	Engagement	8.467	2	4.233	4.363	.023	.244
	Productivity	5.400	2	2.700	2.183	.132	.139
Error	Engagement	26.200	27	.970			
	Productivity	33.400	27	1.237			
Total	Engagement	1648.000	30				
	Productivity	1594.000	30				
Corrected Total	Engagement	34.667	29				
	Productivity	38.800	29				

a. R-Squared =.244 (Adjusted R-Squared =.188)
b. R-Squared =.139 (Adjusted R-Squared =.075)

differences do exist across the groups, we don't yet know (1) which team structure differs from which, (2) whether the effect of team structure is on engagement, productivity, or a combination of both. That is where follow-up analyses are needed.

A significant MANOVA can be followed up by either univariate ANOVAs or a discriminant analysis. If one's focus is on individual dependent variables, conducting simple ANOVA comparisons is a good choice. However, if the relationship between dependent variables is a research interest, discriminant analysis can be a powerful complementary tool. In most cases, univariate ANOVAs are sufficient to understand how each dependent variable individually contributes to the significant MANOVA effect. For simplicity reasons, we will only look at the univariate ANOVA information as the follow-up analysis in this example, which can be found in Table 13.4.

Table 13.5 Contrast Results

TeamStructure Simple Contrast[a]			*Dependent Variable*	
			Engagement	*Productivity*
Level 2 vs. Level 1	Contrast Estimate		.600	.000
	Hypothesized Value		0	0
	Difference (Estimate – Hypothesized)		.600	.000
	Std. Error		.441	.497
	Sig.		.184	1.000
	95% Confidence Interval for Difference	Lower Bound	−.304	−1.021
		Upper Bound	1.504	1.021
Level 3 vs. Level 1	Contrast Estimate		1.300	.900
	Hypothesized Value		0	0
	Difference (Estimate – Hypothesized)		1.300	.900
	Std. Error		.441	.497
	Sig.		.006	.082
	95% Confidence Interval for Difference	Lower Bound	.396	−.121
		Upper Bound	2.204	1.921

a. Reference category = 1

Follow-Up Analysis

The row on *TeamStructure* and the following ANOVA summary will be our focus in Table 13.4. The ANOVA results indicate there are significant differences in engagement (p =.02) but not in productivity across groups (p = .13). This tells us that we need to further look into which group differs from which on employee engagement. To achieve this, we can examine the results from either the Contrasts or the Post-Hoc analysis.

Using the simple contrast method, SPSS compares all the other groups to the reference group. The first group has been used as the default reference group, which is the Virtual team structure in our example. By paying attention to the p-values in the table (Table 13.5), there is only one significant result, that is, the difference in engagement between Hybrid and Virtual team structures (Level 3 vs. Level 1, $p < .01$). You may wonder whether the difference between F2F (group 2) and Hybrid (group 3) is significant because such information is not available in the Contrasts table. Well, we will need to change the reference group in the setting to have that information. In other words, the decision as to which group is best suited to serve as the reference group needs to come from our hypotheses.

The Post-Hoc analysis interface in SPSS for MANOVA is almost identical to what it is for ANOVA. As the number of observations in each group is the same and equal variances are assumed, we can choose Tukey as the post-hoc test. The results are very similar to what we had in the simple contrasts. The only significant difference is in engagement between Hybrid and Virtual team structures (p = .02). In our example, we ended up getting almost the same results using Contrasts and Post-hoc. What is the difference between them, and which one should we choose in a research study? That really depends on our research goals and hypotheses. Use post-hoc tests, shown in Table 13.6, when we have no specific hypotheses and want to explore all possible group comparisons after finding a significant MANOVA result. On the other hand, we use contrasts when we have a strong hunch about where the group differences lie because we are consciously choosing a reference group and comparing it to all the other groups.

Table 13.6 Post-Hoc Test Results

Tukey HSD							
Dependent Variable	*(I) TeamStructure*	*(J) TeamStructure*	*Mean Difference (I-J)*	*Std. Error*	*Sig.*	*95% Confidence Interval*	
						Lower Bound	*Upper Bound*
Engagement	Virtual	F2F	−.60	.441	.374	−1.69	.49
		Hybrid	−1.30*	.441	.017	−2.39	−.21
	F2F	Virtual	.60	.441	.374	−.49	1.69
		Hybrid	−.70	.441	.268	−1.79	.39
	Hybrid	Virtual	1.30*	.441	.017	.21	2.39
		F2F	.70	.441	.268	−.39	1.79
Productivity	Virtual	F2F	.00	.497	1.000	−1.23	1.23
		Hybrid	−.90	.497	.186	−2.13	.33
	F2F	Virtual	.00	.497	1.000	−1.23	1.23
		Hybrid	−.90	.497	.186	−2.13	.33
	Hybrid	Virtual	.90	.497	.186	−.33	2.13
		F2F	.90	.497	.186	−.33	2.13

*. The mean difference is significant at the.05 level.
Based on observed means.
The error term is Mean Square(Error) = 1.237.

Test your understanding: 13.2: What is the difference between using Contrasts and Post-Hoc as the follow-up analysis in MANOVA?

Reporting

There are generally two steps in reporting MANOVA results. First, we report the multivariate effect of the independent variable on the dependent variables by specifying the test statistic used and reporting the *p*-value and the effect size. If multivariate effect is insignificant, the reporting stops here, as the overall insignificant result indicates that the combination of dependent variables is not significantly influenced by the independent variable. Upon a significant multivariate effect, we then move to step two by reporting the follow-up analyses. For instance, we can report the MANOVA results of the example as follow:

> The MANOVA revealed a significant multivariate effect of team structure on employee performance, Wilks' lambda $\lambda = .69$, $F(4, 52) = 2.71$, $p < .05$, $\eta^2 = .17$. Separate follow-up univariate ANOVAs revealed significant effects for *employee engagement*: $F(2, 27) = 4.36$, $p < .05$, $\eta^2 = .24$, but not for *employee productivity*: $F(2, 27) = 2.18$, $p = .13$, $\eta^2 = .14$. Post-hoc Tukey's HSD tests showed that *Hybrid* teams scored significantly higher in *employee engagement* than *Virtual* teams ($p < .05$). No significant differences were observed regarding *employee productivity*.

Recall that the Mikkelson et al. (2011) study utilized MANOVA to compare genetically related siblings to nongenetically related siblings on their relational maintenance behaviors: positivity, assurances, openness, networks, and tasks. Just like in our example above, the study followed a two-step structure, first reporting the overall MANOVA results and then proceeding to the univariate analyses:

> The MANOVA produced significant multivariate main effects for sibling type only, λ =.73, F(5, 381) = 5.13, p <.001, partial η^2 =.06. Univariate analyses for sibling type produced significant main effects for positivity, F(5, 385) = 2.42, p =.035, partial η^2 =.03; openness, F(5, 385) = 4.20, p =.001, partial η^2 =.05; assurances, F(5, 385) = 6.68, p <.001, partial η^2 =.08; networks, F(5, 385) = 13.16, p <.001, partial η^2 =.15; and tasks, F(5, 385) = 8.26, p <.001, partial η^2 =.06" (p. 264).

Summary

In this chapter, we explored the theoretical and practical aspects of conducting a MANOVA. You now understand why MANOVA is often preferred over multiple ANOVAs and what test statistics to choose when performing a MANOVA. You have also learned about the differences between two follow-up analyses and how to appropriately report MANOVA results. At this point, you may try to search for some real-world communication studies that have conducted ANOVA and MANOVA to enhance your mastery of the content. However, you might be rather disappointed in this endeavor, as few studies have directly used them. Instead, their variations, called ANCOVA and MANCOVA, are more popular in social science research. We will delve into the reason for that in the next chapter, but you need to know that what you have learned about ANOVA and MANOVA is fundamental to grasp the concepts in ANCOVA and MANCOVA.

References

Mikkelson, A. C., Myers, S. A., & Hannawa, A. F. (2011). The differential use of relational maintenance behaviors in adult sibling relationships. *Communication Studies*, *62*(3), 258–271. https://doi.org/10.1080/10510974.2011.555490

Stevens, J. P. (1980). Power of the multivariate analysis of variance tests. *Psychological Bulletin*, *88*, 728–737.

Chapter 14

ANCOVA and MANCOVA

What Will I Learn in This Chapter?

If Darth Vader were your teacher for statistics, the classroom culture would be very different. Hypothesis testing now has real consequences, and you are too afraid to tell him that you are only 95% sure of your findings. Out of fear, some students would push themselves to study harder, while others might give up entirely, believing it hopeless to meet Vader's standards. You might expect that hardworking students will perform better on the final exam, yet in this class, everyone, regardless of effort, scores surprisingly low. Does that mean hard work doesn't matter? That conclusion seems farfetched. A more likely explanation is that the intimidating presence of the instructor affected students' academic performance. If we could account for the effect of Vader's intimidation, we might find that students who studied harder did, in fact, achieve higher scores.

DOI: 10.4324/9781003500117-14

This chapter will explore why controlling for variables, like an instructor's intimidation level, is crucial in statistical analysis and how doing so can lead to more accurate conclusions. More specifically, we will cover:

1) Covariates
2) ANCOVA and MANCOVA
3) Performing ANCOVA and MANCOVA in SPSS
4) Interpreting ANCOVA and MANCOVA results

What Are the Covariates and Why Controlling for Them Matters?

Recall that ANOVA is useful in detecting group differences in the dependent variable. A one-way ANOVA essentially tests the effect of the independent variable on the dependent variable. Say you are interested in understanding the weight loss progress among three groups of people based on their daily calorie intake: (1) individuals consuming excessive calories, (2) individuals consuming moderate calories, and (3) individuals consuming insufficient calories. Conducting an ANOVA here will give us answers on whether those who intake less calories experience greater weight loss. However, we know weight loss is also influenced by people's daily activity level. There is a chance those who overeat may also burn a lot of calories and end up in calorie deficit. Therefore, controlling for the effects of calories burnt will give us a better understanding of the effects of calorie intake on weight loss. In this example, calories burnt is called a **covariate** as it is not our primary interest in the study but has an influence on the dependent variable. There are two major benefits when including covariates in ANOVA (Field, 2009):

- **Reducing error variance**: Recall that in ANOVA, we ideally want to have more variance in the dependent variable explained by our model. Now we can explain more variance and thus reduce the noise by including covariates in our model because we know the covariates account for some of the variance in the dependent variable. In simple terms, our overall model is more powerful with covariates in it.
- **Elimination of confounds:** By controlling for the influence of covariates, ANCOVA isolates the unique contribution of the IV on the DV. So, our overall model is more accurate with covariates (useful input) in it.

Including covariates in an ANOVA renders it an **ANCOVA (analysis of covariance)**. These benefits are the primary reason why ANCOVA is much more popular than ANOVA in social scientific research.

<u>**Test your understanding 14.1:**</u> What are the benefits of including covariates in ANOVA?

ANCOVA Assumptions

ANCOVA shares the same assumptions as ANOVA, with two additional requirements. As mentioned earlier, the first advantage of using ANCOVA is that including covariates helps explain some of the error variance. However, this is no longer true when the covariate systematically related to the levels of the independent variable. When the covariate and the independent variable are not independent from each other, their combined influence on the dependent variable can confound the results. This is because ANCOVA assumes the covariate's relationship with the dependent variable is consistent across all levels of the independent variable. When this is violated, adding the covariate may be detrimental to the model because part of the observed effect could be due to systematic differences in the covariate. Thus, the first additional assumption requires the independence of the covariate and treatment effect. We can tackle this issue by examining whether experimental groups differ on the covariate before we run the ANCOVA. And a significant difference result would signal that this assumption is violated.

The second additional assumption is called homogeneity of regression slopes. This assumption requires the relationship (regression slopes) between the covariate and the dependent variable to be similar across all levels of the independent variable. This is important because we are examining the overall relationship between the covariate and the dependent variable in an ANCOVA. When this assumption is violated, the overall regression model fails to accurately represent all groups.

Test your understanding 14.2: What are the additional assumptions to consider when adding covariates to ANOVA?

Performing ANCOVA in SPSS

Let's use the data in Table 14.1 as an example for running an ANCOVA. Again, we are using a simplified data set to demonstrate the process of performing an ANCOVA. Having at least 20 data points in each group is crucial when ANCOVA is performed for a research project. Here, we have data on exam scores and study hours across three different classes. Our goal is to test whether students in these classes significantly differ on exam scores after controlling for study hours.

Let's first perform an ANOVA as a contrast to ANCOVA results. As with the factorial ANOVA, start your ANCOVA as shown in Figure 14.1 by going to:

1. Analyze
2. General Linear Model
3. Univariate
4. As we are only running ANOVA now, we will need to move Score to Dependent Variable and Class to Fixed Factor(s). We will also go to Options and select Descriptive statistics, Estimates of effect size, and Homogeneity tests. Please refer to the chapter on ANOVA for these specification decisions.
5. For post-hoc analysis, we opt for Tukey, as indicated in Figure 14.2, because we have equal numbers in our groups.

Table 14.1 Exam Score across Three Classes

Class 1		*Class 2*		*Class 3*	
Exam Score	*Study Hours*	*Exam Score*	*Study Hours*	*Exam Score*	*Study Hours*
84	4	81	4	65	4
89	5	71	4	72	4
91	4	73	3	83	3
86	5	84	4	70	4
95	4	91	5	53	5
88	5	69	3	61	3
90	5	75	4	48	4

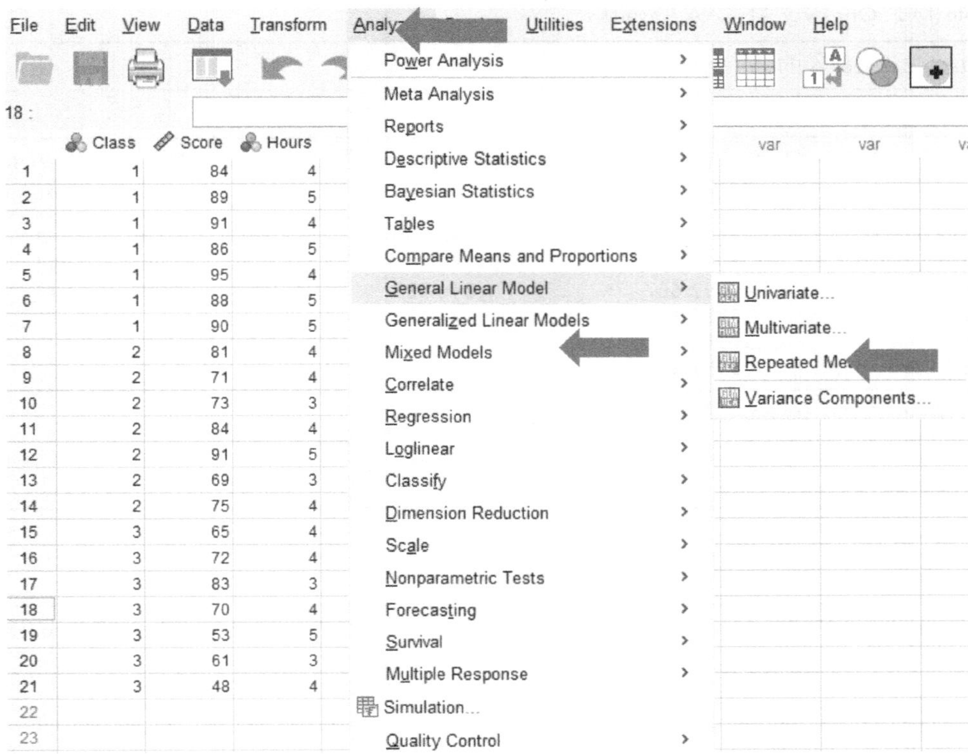

Figure 14.1 Data entry and find GLM univariate.

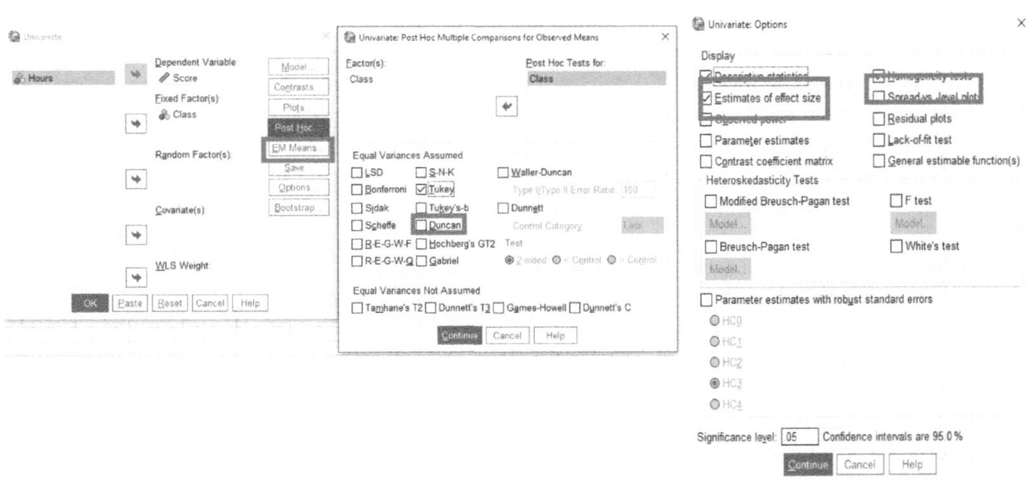

Figure 14.2 ANOVA specification.

Table 14.2 One-way ANOVA Results

Tests of Between-Subjects Effects

Dependent Variable: Score

Source	*Type III Sum of Squares*	*df*	*Mean Square*	*F*	*Sig*	*Partial Eta-Squared*
Corrected Model	2092.667[a]	2	1046.333	14.497	<.001	.617
Intercept	124817.190	1	124817.190	1729.378	<.001	.990
Class	2092.667	2	1046.333	14.497	<.001	.617
Error	1299.143	18	72.175			
Total	128209.000	21				
Corrected Total	3391.810	20				

[a] R-Squared = .617 (Adjusted R-Squared = .574)

Multiple Comparisons

DependentVariable: Score

Tukey HSD

(I) Class	*(J) Class*	*Mean Difference (I-J)*	*Std. Error*	*Sig.*	*95% Confidence Interval*	
					Lower Bound	*Upper Bound*
CLASS 1	CLASS 2	11.29	4.541	.057	−.30	22.88
	CLASS 3	24.43*	4.541	<.001	12.84	36.02
CLASS 2	CLASS 1	−11.29	4.541	.057	−22.88	.30
	CLASS 3	13.14*	4.541	.025	1.55	24.73
CLASS 3	CLASS 1	−24.43*	4.541	<.001	−36.02	−12.84
	CLASS 2	−13.14*	4.541	.025	−24.73	−1.55

We have the results from the one-way ANOVA test in Table 14.2. For those who are curious, you would get the same results if you followed the instructions in Chapter 13 by going to Analyze Compare means and proportions One-way ANOVA. Based on the results, we can conclude there were significant differences in exam scores between different classes, $F\,(2, 18) = 14.\,50$, $p <.001$, $\eta^2 =.62$. The post-hoc results indicated that Class 1 scored significantly higher than Class 2 ($p <.05$) and Class 3 ($p <.001$). Now, let's include the covariate to make it ANCOVA to see whether the results are any different.

Running ANCOVA from here is extremely easy.

1. Follow the same procedure as demonstrated in Figure 14.1.
2. When on the Univariate interface, we also move Hours to Covariate(s) to make it ANCOVA.
3. Specifying the post-hoc test is a little different for ANCOVA because traditional post-hoc procedures use raw group means while ANCOVA essentially analyzes adjusted means. As indicated in Figure 14.3, the Post-Hoc button would become unavailable once you add covariates to the model. However, one can get around this by going to EM Means (Estimated Marginal Means) and selecting the factor that is being examined. Then check Compare Main Effects and choose Bonferroni, which compares the adjusted means.

As shown in Table 14.3, the ANCOVA results are very similar to what we had from the ANOVA test. We can report the ANCOVA results as follows: There were significant differences

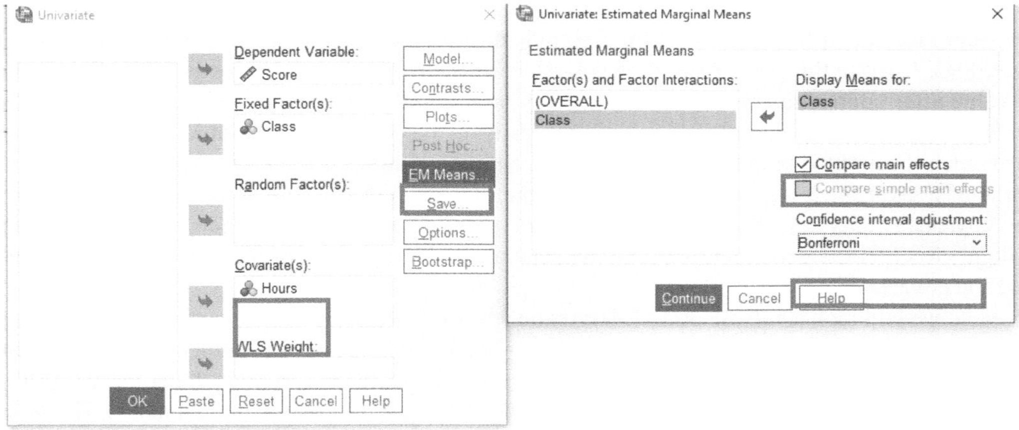

Figure 14.3 Adding a covariate and specifying the post-hoc test.

Table 14.3 ANCOVA Results

Tests of Between-Subjects Effects

Dependent Variable: Score

Source	*Type III Sum of Squares*	*df*	*Mean Square*	*F*	*Sig.*	*Partial Eta-Squared*
Corrected Model	2093.658[a]	3	697.886	9.139	<.001	.617
Intercept	2679.377	1	2679.377	35.088	<.001	.674
Hours	.992	1	.992	.013	.911	.001
Class	1764.660	2	882.330	11.555	<.001	.576
Error	1298.151	17	76.362			
Total	128209.000	21				
Corrected Total	3391.810	20				

a. R-Squared = .617 (Adjusted R-Squared = .550)

Pairwise Comparisons

DependentVariable: Score

					95% Confidence Interval for Difference[b]	
(I) Class	*(J) Class*	*Mean Difference (I-J)*	*Std. Error*	*Sig.*[b]	*Lower Bound*	*Upper Bound*
CLASS 1	CLASS 2	11.547	5.202	.121	−2.265	25.358
	CLASS 3	24.690*	5.202	<.001	10.878	38.501
CLASS 2	CLASS 1	−11.547	5.202	.121	−25.358	2.265
	CLASS 3	13.143*	4.671	.036	.742	25.544
CLASS 3	CLASS 1	−24.690*	5.202	<.001	−38.501	−10.878
	CLASS 2	−13.143*	4.671	.036	−25.544	−.742

Based on estimated marginal means

*. The mean difference is significant at the 0.05 level.

b. Adjustment for multiple comparisons: Bonferroni.

in exam scores between different classes, $F(2, 17) = 11.56$, $p < .001$, $\eta^2 = .58$. The Bonferroni Post-hoc test found that Class 1 scored significantly higher than Class 2 ($p < .05$) and Class 3 ($p < .001$). Despite the similarities between the results, we do notice there was a decrease in the F-value after controlling for the effect of study hours on the exam score. This is often the case when the covariate accounts for some variability in the dependent variable that might have previously been attributed to the independent variable. A lower F-value in ANCOVA indicates some of the variance previously attributed to the independent variable is now accounted for by the covariate. This adjustment results in a more precise estimate of the independent variable's unique effect on the dependent variable by separating the influence of the independent variable and the covariates.

Next, we will look at the use of ANCOVA with a communication research example. In Sangalang and Bloomfield's (2018) study examining how different story types might influence the perception of climate change, ANCOVA was performed to answer the research questions. Let's focus on RQ_1, which explored which story type can produce the highest levels of perceived effectiveness. The independent variable had five levels, as the study was designed to include five distinctive story types. The controlling variables (covariates) were sex, race, religious affiliation, and highest education level. First, the study reported the main ANCOVA results, which revealed significant differences between story types, $F(4, 566) = 4.90$, $p < .01$. Then, the Bonferroni Post-hoc test revealed that the Type III story type had significantly higher levels of perceived effectiveness than Type II and Type IV.

MANCOVA

Like the relationship between ANOVA and ANCOVA, MANCOVA (**Multivariate analysis of covariance**) is MANOVA with a covariate(s). It might be hard to believe, but you already know how to perform a MANCOVA by now. This is because running MANCOVA is no more than combining what we have learned about MANOVA and ANCOVA. In addition to sharing all the assumptions of MANOVA, MANCOVA also requires (1) independence between covariates and treatment effects and (2) homogeneity of regression slopes, almost identical to the differences between ANCOVA and ANOVA.

In Table 14.4, study motivation, another dependent variable, has been added to the ANCOVA data. Now with two dependent variables and one covariate, we need to perform a MANCOVA to understand whether exam score and study motivation differ significantly across three classes while controlling for study hours.

Table 14.4 MANCOVA Data on Exam Scores and Study Motivation Differences across Classes

Class 1			*Class 2*			*Class 3*		
Exam Score	*Motivation*	*Study Hours*	*Exam Score*	*Motivation*	*Study Hours*	*Exam Score*	*Motivation*	*Study Hours*
84	7	4	81	8	4	65	6	4
89	7	5	71	8	4	72	8	4
91	8	4	73	9	3	83	7	3
86	9	5	84	7	4	70	8	4
95	8	4	91	8	5	53	5	5
88	7	5	69	6	3	61	5	3
90	10	5	75	7	4	48	6	4

In SPSS, running a MANCOVA is almost identical to running an ANCOVA, with the only difference being that we have two dependent variables instead of one. We will take the following steps to perform a MANCOVA.

1. Go to Analyze
2. General Linear Model
3. Multivariate
4. Move both Score and Motivation to Dependent Variables
5. Add Hours to the Covariate(s)
6. Click on Options and select Descriptive statistics, Estimates of effect size, and Homogeneity tests
7. Go to EM Means
8. Display means for Class
9. Check Compare main effects and choose Bonferroni
10. Then click OK to get to the MANCOVA results

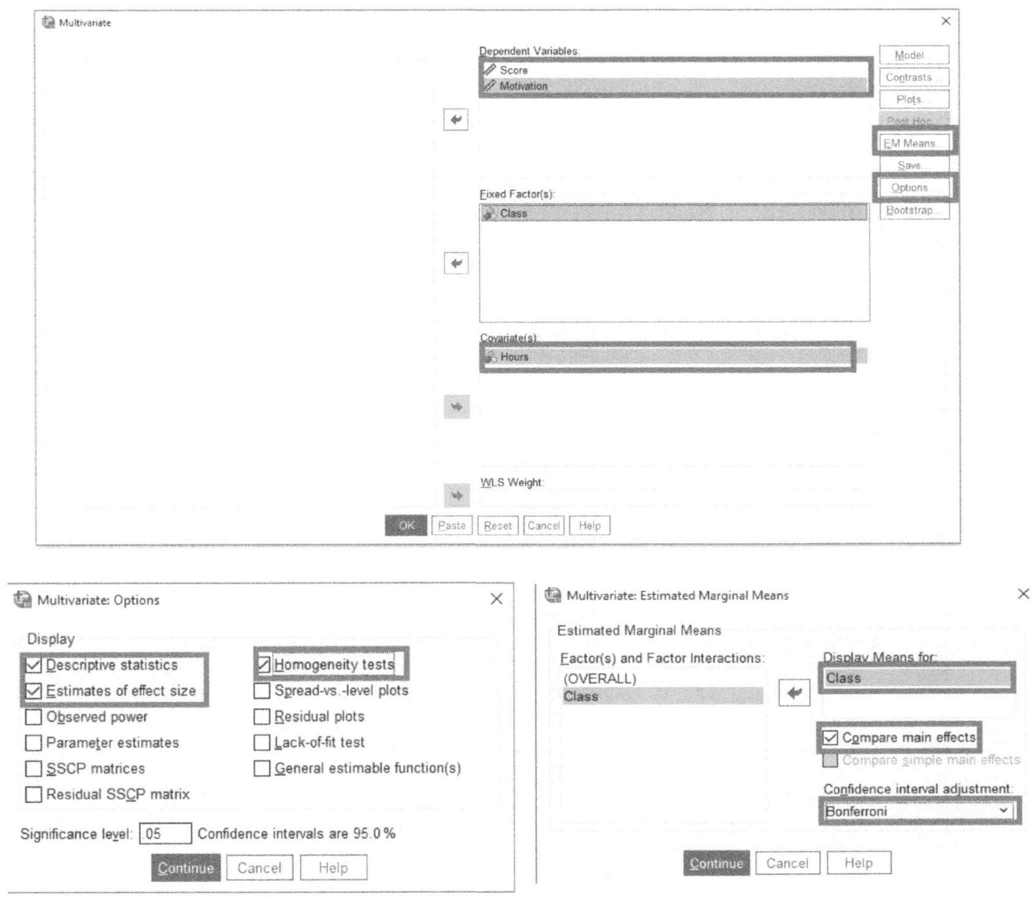

Figure 14.4 Running MANCOVA.

We will skip the explanation of the results of the homogeneity tests, as they are the same as what we explained in the previous chapter. The first thing we want to pay attention to is the overall multivariate test results in Table 14.5, which evidenced significant findings: Wilks' λ =.41, F (4, 32) = 4.47, p <.01. partial η^2 = 0.36. With that encouraging news, we then look at the univariate ANOVAs in Table 14.6 to see which specific dependent variable(s) differs across groups. It is evident that significant differences existed in exam score (p <.001) but not in study motivation (p >.05). The post-hoc results in Table 14.7 further showed us that Class 1 (p <.001). and Class 2 (p <.05) scored significantly higher than Class 3.

Table 14.5 MANCOVA Multivariate Test Results

	Value	*F*	*Hypothesis df*	*Error df*	*Sig.*	*Partial Eta-Squared*
Pillai's trace	.604	3.682	4.000	34.000	.014	.302
Wilks' lambda	.412	4.470	4.000	32.000	.006	.358
Hotelling's trace	1.391	5.216	4.000	30.000	.003	.410
Roy's largest root	1.362	11.581	2.000	17.000	<.001	.577

Table 14.6 MANCOVA Univariate Test Results

Source	*Dependent Variable*	*Sum of Squares*	*df*	*Mean Square*	*F*	*Sig.*	*Partial Eta-Squared*
Contrast	Score	1764.660	2	882.330	11.555	<.001	.576
	Motivation	7.458	2	3.729	2.722	.094	.243
Error	Score	1298.151	17	76.362			
	Motivation	23.294	17	1.370			

Table 14.7 MANCOVA Post-Hoc Results

Dependent Variable	*(I) Class*	*(J) Class*	*Mean Difference (I-J)*	*Std. Error*	*Sig.*[b]	*95% Confidence Interval for Difference*[b]	
						Lower Bound	*Upper Bound*
Score	CLASS 1	CLASS 2	11.547	5.202	.121	−2.265	25.358
		CLASS 3	24.690*	5.202	<.001	10.878	38.501
	CLASS 2	CLASS 1	−11.547	5.202	.121	−25.358	2.265
		CLASS 3	13.143*	4.671	.036	.742	25.544
	CLASS 3	CLASS 1	−24.690*	5.202	<.001	−38.501	−10.878
		CLASS 2	−13.143*	4.671	.036	−25.544	−.742
Motivation	CLASS 1	CLASS 2	.332	.697	1.000	−1.518	2.183
		CLASS 3	1.475	.697	.148	−.375	3.325
	CLASS 2	CLASS 1	−.332	.697	1.000	−2.183	1.518
		CLASS 3	1.143	.626	.256	−.518	2.804
	CLASS 3	CLASS 1	−1.475	.697	.148	−3.325	.375
		CLASS 2	−1.143	.626	.256	−2.804	.518

*. The mean difference is significant at the .05 level.
b. Adjustment for multiple comparisons: Bonferroni.
Based on estimated marginal means

We can report the MANCOVA results as follows: The MANCOVA revealed a significant multivariate effect, Wilks' lambda λ =.41, F (4, 32) = 4.47, p <.01, partial η^2 = 0.36. Follow-up univariate ANOVAs revealed significant differences in exam score – F (2, 17) = 11.56, p <.001, Partial η^2 =.58, but not for study motivation – F (2, 17) = 2.72, p >.05, Partial η^2 =.24. The Bonferroni test results showed that Class 1's exam score was significantly higher than Class 2 (p <.05) and Class 3 (p <.001).

Next, let's see how MANCOVA is used with a communication research example. In Zeng et al. (2020)'s study, the authors hypothesized that when unethical events happen in the workplace, both the unethical nature of the event and the leader's integrity will have an influence on employees' likelihood of reporting it to organizational leaders and to outside agencies. There are two independent variables (unethicalness and leader integrity) and two dependent variables (employees' internal and external voice intentions). As organizational tenure and holding a managerial position have been shown in previous studies to influence employees' voice intentions, this study included them as covariates. Here, MANCOVA is a great solution to include all the variables in one test to reduce the chance of Type I errors.

The study results section started with interpreting the results of the tests of homogeneity. Then, it proceeded to report the main MANCOVA results: Wilks' λ =.97, F (2, 286) = 4.39, p =.01, η^2 =.03. This significant finding indicated that there were significant differences in employees' voice intentions. Univariate post-hoc analyses were then performed to pinpoint where exactly the differences were. The unethicalness of the event had no significant effects on employee's internal voice intention [F(1, 291) = 3.11, p =.08] or external voice intention [F(1, 291) =.46, p =.50]. Leader integrity had significant influences on employees' internal voice intention (F(3, 289) = 12.87, p <.001), but not on external voice intention [F(1, 291) =.41, p =.52]. After reporting these results, the study was able to declare what hypotheses had been supported or not supported. For more communication studies that have utilized MANCOVA, you can go to Croucher et al. (2014) and Nan and Madden (2012).

Summary

In this chapter, we have further expanded our understanding of the family of ANOVAs by introducing covariates to the picture. By controlling for the effects of the covariates, ANCOVA and MANCOVA can enhance the precision of results and reduce error variance. Two additional assumptions require our attention: independence of the covariate and treatment effect, and homogeneity of regression slopes. With previous knowledge on ANOVA and MANOVA, it is easy for us to run ANCOVA and MANCOVA on SPSS, as it is a very similar procedure with mostly one additional step: adding the covariate(s). You have also learned how to appropriately interpret and report ANCOVA and MANOVA results. The failure to include relevant covariates can weaken the validity of an analysis, making it a common point of criticism in rigorous statistical research. In summary, adding covariates allows for a more precise estimation of the true effect of the independent variable(s) on the dependent variable and is generally recommended in social scientific research.

References

Croucher, S. M., Parrott, K., Zeng, C., & Gomez, O. (2014). A cross-cultural analysis of organizational dissent and workplace freedom in five European economies. *Communication Studies*, *65*(3), 298–313. https://doi.org/10.1080/10510974.2013.811430

Field, A. (2009). *Discovering statistics using SPSS* (3rd ed.). Sage.

Nan, X., & Madden, K. (2012). HPV vaccine information in the blogosphere: How positive and negative blogs influence vaccine-related risk perceptions, attitudes, and behavioral intentions. *Health Communication*, *27*(8), 829–836. https://doi.org/10.1080/10410236.2012.661348

Sangalang, A., & Bloomfield, E. F. (2018). Mother goose and mother nature: Designing stories to communicate information about climate change. *Communication Studies*, *69*(5), 583–604. https://doi.org/10.1080/10510974.2018.1489872

Zeng, C., Kelly, S., & Goke, R. (2020). Exploring the impacts of leader integrity and ethics on upward dissent and whistleblowing intentions. *Communication Reports*, *33*(2), 82–94. https://doi.org/10.1080/08934215.2020.1761413

Chapter 15

Repeated Measures

What Will I Learn in This Chapter?

For those who know little about fashion, deciding what color to wear can be a huge challenge. It would be great if we lived in a "Fashion Multiverse Simulator" where we could try out different colors simultaneously and then instantly find out which one we like the most. Sadly, we live in no such world and need to laboriously try them out one by one. Now, instead of comparing different groups of people, we're looking at how we feel in different outfits at different times. It is tempting to apply one-way ANOVA here again because we are essentially comparing means across different conditions. However, the conditions in this case are not independent events, as the same person is being measured multiple times. This violates one

DOI: 10.4324/9781003500117-15

key assumption in ANOVA: the independence of observations. This chapter is devoted to handling situations like this, as we explore the ideas behind repeated-measures design and its implications by covering:

1) The rationale of repeated-measures design
2) The assumption of sphericity
3) Running repeated-measures ANOVA on SPSS
4) Interpreting and reporting repeated-measures ANOVA results

Why and How Repeated-Measures Design Is Different?

In previous ANOVA-related chapters, we have mostly discussed testing for differences across groups. In this chapter, we will shift our focus to studies that track the same group of people over a period of time. In the previous example, we are tracking our mood by wearing different colors of clothes on different days. Now, can we still use ANOVA to test the mean differences between different time points? The answer is yes and no. Exposing the same participants to multiple conditions over time is known as a **repeated-measures design.** This is to distinguish from an **independent groups design,** where studies have separate groups of participants for each condition, with no overlap. In Chapter 12, we discussed that one key assumption of ANOVA is the independence of observations, which requires that the data points in the study are unrelated to one another. Violation of this assumption will render the *F*-test in ANOVA inaccurate. In repeated-measures designs, we are tracking the same participants over time, which means our data is bound to be interrelated and will surely violate the independence of observations assumption. To tackle this issue, we need to use repeated-measures ANOVA that accounts for the correlation between repeated measurements.

Test your understanding 15.1: What independent groups ANOVA assumption is violated when we have a repeated-measures design?

In independent groups ANOVA, one important assumption is called homogeneity of variance, which requires that the variances of the dependent variable are equal across all groups. Although we have only one group of participants in repeated-measures designs, the same idea that variances of the dependent variable are equal across all conditions also applies. This is called the assumption of **sphericity**. When performing repeated-measures ANOVA, SPSS automatically checks this assumption by running **Mauchly's test.** Just like the Levene's test for ANOVA and the Box's test for MANOVA, a significant Mauchly's test statistic indicates that the differences between the variances of differences are significant, and thus the assumption of sphericity is violated. There is no need to panic when this happens, as SPSS automatically produces two corrections based upon the estimates. They are to be used under different conditions:

- **Greenhouse–Geisser correction:** More conservative, used when sphericity is moderately violated (e.g., the correction factor is far below 1).
- **Huynh–Feldt correction:** Less conservative, used when the violation is minor (e.g., when the correction factor is closer to 1).

In Chapter 12, we mentioned that independent ANOVA has the following types of variances: the total sum of squares (SS_T), the model sum of squares (SS_M), and the residual sum of squares (SS_R). The F-test is calculated with the following formula:

$$F = \frac{\mathrm{MS_M}}{\mathrm{MS_R}} = \frac{SS_M \,/\, df_M}{SS_R \,/\, df_R}$$

This F-test formula also applies to repeated-measures ANOVA that has the same categories of variances. However, SS_M refers to the sum of squares between groups in independent groups ANOVA, whereas SS_M in repeated-measures ANOVA is calculated by SS_W (Sum of Squares Within Subjects) subtracting SS_R (Sum of Squares for Residuals/Error). This key difference exists because we are interested in the variation within a person instead of a group of people in repeated-measures designs. Let's use the following example to help illustrate the key points in running repeated-measures ANOVA on SPSS.

Test your understanding 15.2: What are the similarities between running a repeated-measures ANOVA and an independent groups ANOVA?

Repeated-Measures ANOVA in SPSS

Imagine we have devised a communication training program to help improve students' public speaking skills. Their skills are measured on a scale of 1 to 10, and we have collected data over a span of three weeks, as presented in Table 15.1. (Again, in the real world, have at least 20 data points per cell before analyzing data with any type of ANOVA, i.e., here, 20 data points each week).

First, we can find the repeated measures by going to (see Figure 15.1):

1. Analyze
2. General Linear Model
3. Repeated Measures
4. Next, we will need to define the name of the within-subject factor. In our example, such a factor is the time points, as we have measured students' skills over a period of three weeks. You are encouraged to change the factor's name if you have more variables in the dataset, as you may lose track of which is which. Here, we will just go with the default by using factor1 and define the number of levels as 3 since we have 3 data points (see Figure 15.2).
5. Click Add
6. Click Define

Table 15.1 Longitudinal Data on Public Speaking Training over Three Weeks

	Week 1	*Week 2*	*Week 3*
Participant 1	4	6	7
Participant 2	5	5	6
Participant 3	6	7	9
Participant 4	4	5	8
Participant 5	3	6	9

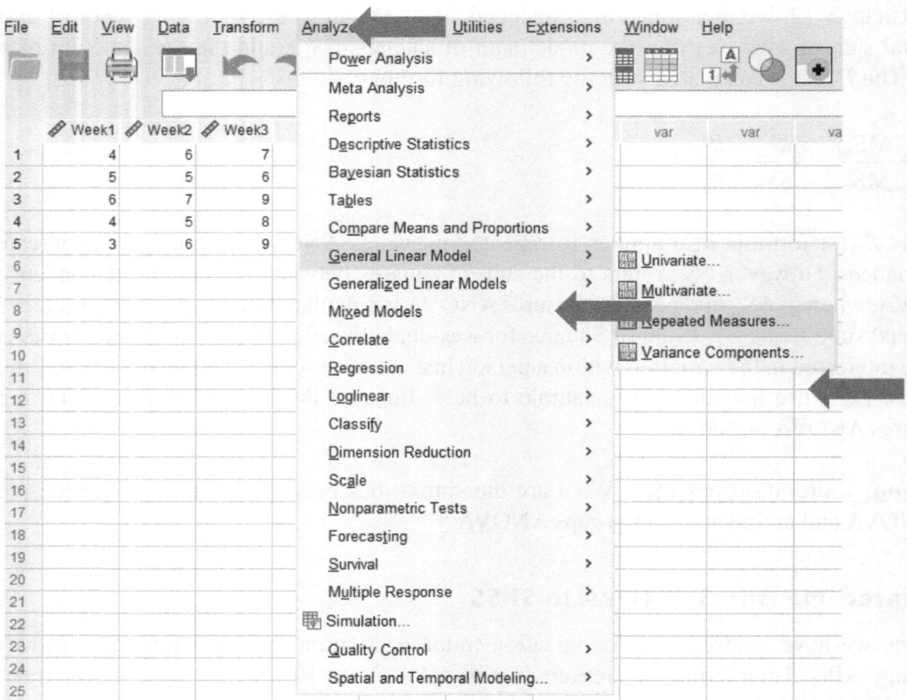

Figure 15.1 Data entry and find repeated-measures ANOVA in SPSS.

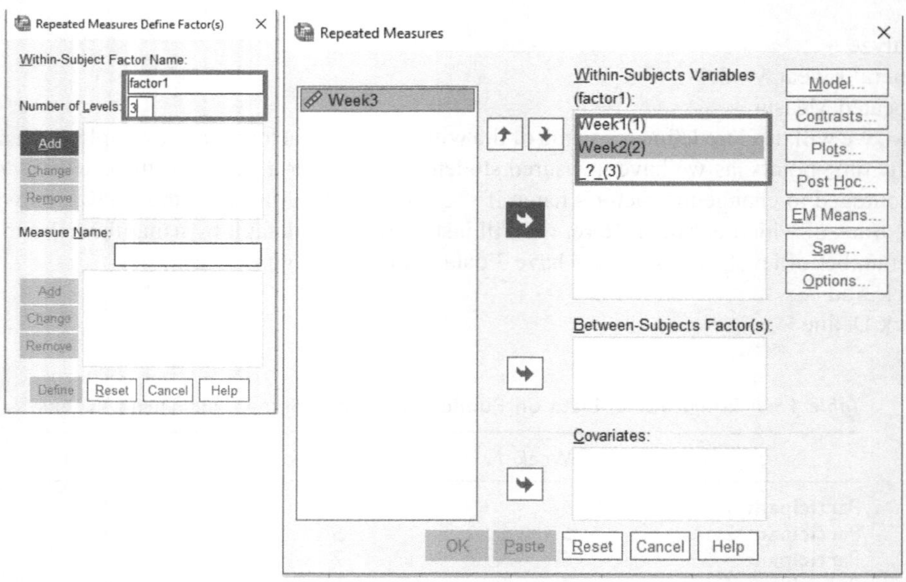

Figure 15.2 Main analysis specification.

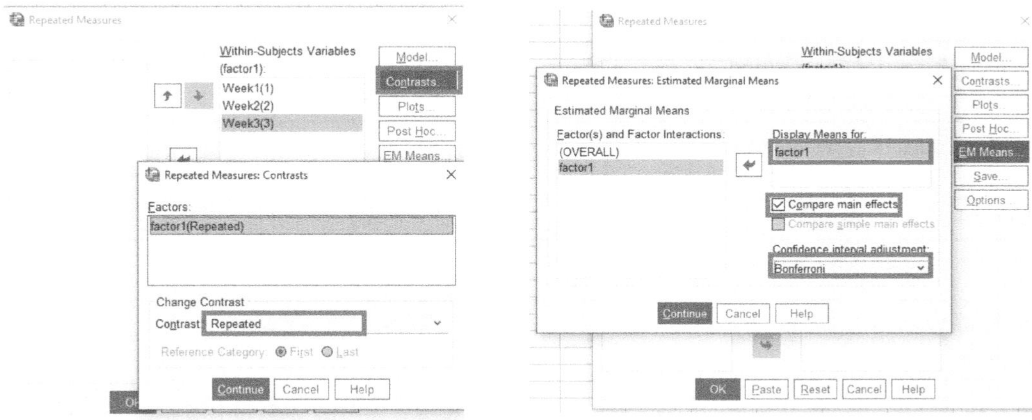

Figure 15.3 Post-hoc tests specification.

You will be taken back to the main interface, where you get to match the data to the defined levels. As our data is already organized in chronological order, we will simply move the weeks to the defined three levels accordingly.

Completing the aforementioned steps allows us to perform the main analysis of the repeated measures. However, extra steps are needed if we want post-hoc information on the differences between each week. Post-hoc tests in SPSS are typically available only for between-subjects factors, not within-subjects factors. Post-hoc comparisons for the repeated measures can still be achieved in two ways. The first option is to use Contrasts to make pairwise comparisons. While we are no stranger to this function at this point, notice that the Contrast method needs to be changed to Repeated to reflect the nature of the data (Figure 15.3). The second way is using Bonferroni correction tests, just like what we did for ANCOVA and MANCOVA. We can click on EM Means and move factor1 to the right side, then click the Compare main effects box and change the confidence interval adjustment to Bonferroni. Finally, we can go to Options and select Descriptive statistics and Estimates of effect size, just like we do for all the other ANOVA tests. Having completed these specifications, we are ready to perform the test.

Interpreting the Results

When it comes to interpreting the SPSS output, the first step is to assess the assumption of sphericity by checking the results of Mauchly's test. As evidenced in Table 15.4, Mauchly's test is nonsignificant ($p > .05$), indicating that the assumption of sphericity is met. This information is very useful for interpreting the overall ANOVA table, which is named Tests of Within-Subjects Effects. This table of repeated-measures ANOVA can be interpreted in the same way as independent groups ANOVA, as explained in Chapter 12. You will notice that SPSS produced different rows based on whether the assumption of sphericity is violated. We will go ahead and interpret the first row, as sphericity is assumed in our study based on the results of Mauchly's test. In the case of sphericity violation, we can interpret results from the latter rows, depending on the severity of the violation as mentioned before.

Table 15.2 Mauchly's Test and the Main ANOVA Table

Mauchly's Test of Sphericity[a]
Measure: MEASURE_1

Within-Subjects Effect	*Mauchly's W*	*Approx. Chi-Square*	*df*	*Sig.*	*Greenhouse–Geisser*	*Epsilon*[b] *Huynh–Feldt*	*Lower bound*
factor1	.402	2.736	2	.255	.626	.774	.500

Tests of With in-Subjects Effects
Measure: MEASURE_1

Source		*Type III Sum of Squares*	*df*	*Mean Square*	*F*	*Sig.*	*Partial Eta-Squared*
factor1	Sphericity Assumed	29.200	2	14.600	15.643	.002	.796
	Greenhouse–Geisser	29.200	1.251	23.334	15.643	.009	.796
	Huynh–Feldt	29.200	1.549	18.854	15.643	.005	.796
	Lower bound	29.200	1.000	29.200	15.643	.017	.796
Error(factor1)	Sphericity Assumed	7.467	8	.933			
	Greenhouse–Geisser	7.467	5.006	1.492			
	Huynh–Feldt	7.467	6.195	1.205			
	Lower bound	7.467	4.000	1.867			

Reading from Table 15.2, the sum of squares for factor1 (SS_M) is 29.2, and the sum of squares of error (SS_R) is 7.47. For those who want to try out using the *F*-test formula, you will find that the *F*-value is 15.64, which is what is shown in the ANOVA table. This again confirms that both independent groups ANOVA and repeated-measures ANOVA use the same *F*-test formula.

$$F = \frac{\mathrm{MS_M}}{\mathrm{MS_R}} = \frac{SS_M / df_M}{SS_R / df_R} = \frac{29.2/2}{7.47/8} = \frac{14.6}{0.93} = 15.64$$

Here, SS_M refers to the sum of squares for the within-subjects factor, and SS_R represents the residual sum of squares. Degrees of freedom (*df*) correspond to the respective sums of squares.

With the significant *F*-value, we can conclude that there are significant differences in students' public speaking skills over these three weeks. Next, let's turn to pairwise comparisons to find out where exactly the differences lie. We can find the results from the Contrasts in Table 15.2. It is evident that the improvement from Week 1 to Week 2 approached significance (p =.052) but did not meet the conventional threshold, while the improvement from Week 2 to Week 3 was statistically significant ($p < .05$).

Instead of using a main reference group in the Contrasts table, the Bonferroni Pairwise Comparisons results in Table 15.4 compare all groups to each other. The same conclusion can be drawn from the results here: the difference between Week 1 and Week 2 (p >.05) was non-significant while there was a significant difference between Week 2 and Week 3 (p <.05). We can also tell that the difference between Week 1 and Week 3 was significant (p <.05).

Table 15.3 Tests of Within-Subjects Contrasts

Source	*factor1*	*Type III Sum of Squares*	*df*	*Mean Square*	*F*	*Sig.*	*Partial Eta-Squared*
factor1	Level 1 vs. Level 2	9.800	1	9.800	7.538	.052	.653
	Level 2 vs. Level 3	20.000	1	20.000	20.000	.011	.833
Error(factor1)	Level 1 vs. Level 2	5.200	4	1.300			
	Level 2 vs. Level 3	4.000	4	1.000			

Table 15.4 Pairwise Comparisons

(I) factor1	*(J) factor1*	*Mean Difference (I-J)*	*Std. Error*	*Sig.*	*95% Confidence Interval for Difference*	
					Lower Bound	*Upper Bound*
1	2	−1.400	.510	.155	−3.420	.620
	3	−3.400*	.812	.042	−6.618	−.182
2	1	1.400	.510	.155	−.620	3.420
	3	−2.000*	.447	.033	−3.771	−.229
3	1	3.400*	.812	.042	.182	6.618
	2	2.000*	.447	.033	.229	3.771

Reporting

We can report the findings from the previous example in APA as follows:

> Mauchly's test indicated that the assumption of sphericity had been met, $\chi 2(2) = 2.74$, $p > .05$. The repeated-measures ANOVA revealed a significant effect of public speaking training on students' communication skills, F (2, 8) = 15.64, $p < .01$, $\eta^2 = .80$. Pairwise comparisons using Bonferroni adjustments showed that students in Week 3 scored higher than Week 1 ($p < .05$) and Week 2 ($p < .05$).

Let's look at a communication study that has employed the repeated measures design to accomplish its research goals. In a study looking into the language use in condom negotiation by African American women, Li et al. (2018) counted the frequency of four types of language (I-, you-, we-, and hedging) uttered by the participants in an experiment and wanted to see whether there were differences in the use of the languages. Since this scenario involves measuring the same individual throughout the experiment, running a repeated-measures ANOVA would be most appropriate. The study first reported the overall repeated-measures ANOVA test as Wilks's lambda =.76, F (3, 189) = 19.87, $p < .001$, indicating that there were significant differences in the language use. Then, the Bonferroni correction post-hoc tests revealed that "I-language was used significantly more than you-language ($p = .011$), we-language ($p < .001$), and hedging language ($p < .001$)."

Summary

Repeated-measures design is a very common and powerful approach. We can track changes over time or across conditions within the same individuals, providing a clearer understanding of within-subject effects. This information is extremely useful in assessing the effectiveness of a program over a period of time, such as the example we used in this chapter. However, extra caution is also warranted, as new issues may arise as participants are asked to complete the same questions multiple times. For instance, participants may show improvement due to the effects of practicing on the same questions or experience fatigue due to prolonged testing. Also, keeping the response rates high from the same participants repeatedly can be a huge challenge. In this chapter, we have explored how and why repeated-measures design is different from the independent groups ANOVAs that have been our focus in previous chapters. One new assumption, named the assumption of sphericity, needs to be checked, and we have discussed solutions for scenarios where the assumption is met, slightly violated, and severely violated. Defining and matching the levels of the factors can be somewhat different than other tests from the ANOVA family. Otherwise, running and interpreting the main analysis and post-hoc comparisons should be a rather familiar process at this point in the book.

Reference

Li, Y., Samp, J. A., Coles Cone, V. B., Mercer Kollar, L. M., DiClemente, R. J., & Monahan, J. L. (2018). African American women's language use in response to male partners' condom negotiation tactics. *Communication Studies*, *69*(1), 67–84. https://doi.org/10.1080/10510974.2017.1412335

Chapter 16

Chi-Square (χ^2)

What Will I Learn in This Chapter?

The Empire is interested in knowing to what extent male and female Stormtroopers prefer blue over red bicycles. Recently the Emperor has instructed his marketing teams to conduct surveys of all Stormtroopers to ascertain the bicycle color preference across the galaxy. A total of 400 Stormtroopers were gathered on a cruiser to conduct a marketing analysis. In the ideal Empire, there would be four equal groups: 100 males who like red, 100 males who like blue, 100 females who like blue, and 100 females who like red bicycles. However, this is rarely the case. The marketing team found 135 men preferred the blue bicycles versus 65 men who preferred the red, while 110 women preferred the red versus 90 who preferred the blue. Is this a statistical difference? This question can be answered, but not with tests we have covered thus far in this book. In this case, you will need to conduct a chi-square (χ^2) test of independence. In this chapter, we will explain why a chi-square (χ^2) is needed for this test.

DOI: 10.4324/9781003500117-16

In this chapter, you will be introduced to the chi-square (χ^2) test of independence. Specifically, in this chapter, you will:

1. Understand what the chi-square (χ^2) test of independence is and how it differs from previous statistical tests
2. Learn how to run the chi-square (χ^2) test of independence in SPSS

What Are Chi-Square (χ^2) Tests of Independence?

Looking back to Chapters 8–15, you have been working with parametric data and tests. Look back to Chapter 6, where we outlined the key assumptions of parametric data: interval level data (at least), independence, homogeneity of variance, and normality. Non-parametric tests do not require these assumptions to be met, and often these assumptions are violated. In the case of our bicycle preference analysis, such a statistical analysis violates the assumptions of parametric data because both sex (male and female) and preference (red or blue) are nominal level variables. Thus, we cannot run a parametric test like regression, ANOVA, or a Pearson's correlation. We need to run a non-parametric test.

Test Your Understanding 16.1 – What are the differences between parametric and non-parametric tests?

A **chi-square** (χ^2) is a non-parametric test that compares the observed frequencies of a variable against the expected frequencies to see if there is a statistical difference. A basic principle of the chi-square (χ^2) is that it compares observed frequencies to expected frequencies based on a null hypothesis. For example, if you are doing a taste test of Pepsi, Diet Pepsi, Diet Coke, and Coke Zero among 200 people, probability dictates that you should get 50 people in each category. However, this rarely happens. What is observed is often rather different from what is hypothesized or expected. The statistical extent of this difference is what is calculated by the chi-square (χ^2). The greater the difference between the observed and expected frequencies, the higher the χ^2 value, increasing the likelihood of being statistically significant (p <.05). An example of a chi-square (χ^2) often cited in the news is when surveys report differences in voting preferences by gender. For instance, a study might find that men and women differ significantly in their choice of candidate. This comparison of observed versus expected voting patterns across gender categories is an example of how a χ^2 test can be used to analyze categorical data (i.e., gender vs. candidate choice).

Haskollar and Bagwe (2023), in an analysis of the relationship between student characteristics and intercultural competence, used chi-square (χ^2) to understand the link between race/ethnicity and developmental orientation. They found a statistically significant difference between racial/ethnic groups on developmental orientation: $\chi^2(18) = 32.95$, p <.017. In another study, Shulman and DeAndrea (2014) found participants were more likely to have the same political affiliation as their mothers than their fathers: $\chi^2(1) = 4.21$, p <.05.

How to Run Chi-Square (χ^2) Tests of Independence in SPSS

Running a chi-square (χ^2) in SPSS is relatively simple. The data for this chapter is 681 U.S. participants who were surveyed regarding their perceptions about the COVID-19 vaccine and fears about the spread of the virus in the U.S.A. in 2022. First, as shown in Figure 16.1, open the Chi-Square menu by clicking:

1. Analyze
2. Non-parametric Tests

3. Legacy Dialogs
4. Chi-square

These steps will open the window for the chi-square test shown in Figure 16.2. Within this window:

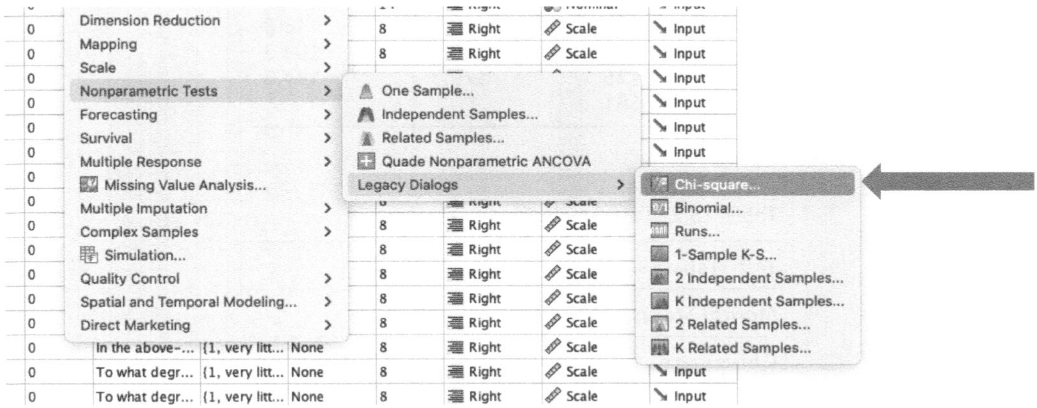

Figure 16.1 Calling up the Chi-square test in SPSS.

5. Click on the first variable you want to analyze and then move it to the "Test Variable List." In this case, we will be looking at the marital status of participants from the United States (nominal variable).
6. Click OK

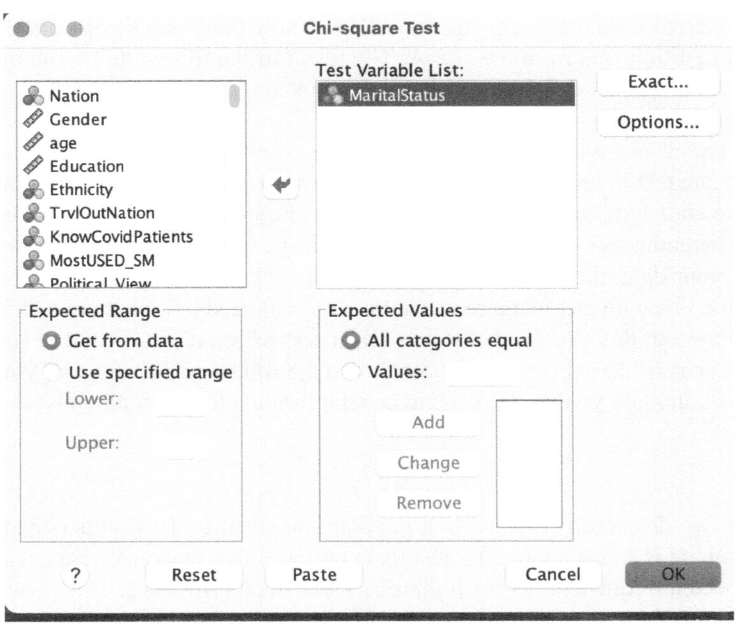

Figure 16.2 Chi-square test.

Table 16.1 Frequencies (Expected vs. Observed)

Marital Status			
	Observed N	*Expected N*	*Residual*
married	270	170.3	99.8
single	319	170.3	148.8
divorced	66	170.3	−104.3
Widowed	26	170.3	−144.3
Total	681		

Table 16.2 Test Statistics

Test Statistics	
Marital Status	
Chi-Square	374.465[a]
df	3
Asymp. Sig.	<.001

[a] 0 cells (0.0%) have expected frequencies less than 5. The minimum expected cell frequency is 170.3.

You will get two outputs, which are important for you to have when writing up your results. Table 16.1 shows the observed versus expected outcome.

The second output, Table 16.2, is the Test Statistics. This output shows the result of the chi-square (χ^2). The results from this chi-square analysis show there is a significant difference in the marital status of participants from the U.S.A. The result of the test could be summarized as such: Results reveal a significant difference in marital status among U.S. participants, $\chi^2(3) = 374.47$, $p < .001$.

Remember Excel? We will note here that Excel can run a chi-square test, but not on raw data. To run a chi-square test in Excel, you must create a new table telling Excel how many counts exist in each group overall and how many you expect to see. From there, you use the function =CHISQ.TEST(). In the parentheses of that formula, you provide the range of cells with the actual frequency distribution of your data, then a comma, then the cells with the expected frequency distribution. This will provide you with a chi-square value that you will need to look up in a chi-square table to interpret. The moral of this story is that if you can access SPSS, use it instead of Excel to run your chi-square test, because Excel alone will not tell you the full story. For those of you out there who really love Excel, though, you're welcome to use the formula to run your analyses there.

Summary

In this chapter, we discussed chi-square, a popular statistical tool for understanding patterns in data. The chi-square is a non-parametric test that compares the observed frequencies of a variable against the expected frequencies to see if there is a statistical difference. This is one of many non-parametric tests out there that you can use. For other tests, talk with your instructor, as there are countless tools you can use ☺.

References

Haskollar, E., & Bagwe, T. K. (2023). Do demographics matter? The relationship between student characteristics and intercultural competence. *Journal of Intercultural Communication Research*, *51*(1), 25–55. https://doi.org/10.1080/17475759.2022.2102059

Shulman, H. C., & DeAndrea, D. C. (2014). Predicting success: Revisiting assumptions about family political socialization. *Communication Monographs*, *81*(3), 386–406. https://doi.org/10.1080/03637751.2014.936478

Chapter 17

Reliability Analysis

What Will I Learn in This Chapter?

Princess Leia once said, "Never underestimate a droid." She knew, as does anyone who has watched *Star Wars,* that droids are courageous, dependable, honest, and reliable. In fact, many of the difficult situations the Rebel Alliance got out of were thanks to droids. We defined reliability in Chapter 2 as the extent to which a research tool, measure/instrument, survey, or process produces consistent results when applied repeatedly under similar conditions. Essentially, reliability is how trustworthy, accurate, and dependable our measures are in producing consistent and stable assessments across samples. In the end, we want our measures to be like R2, C3PO, and BB8 – reliable. This is what we will cover in this chapter.

DOI: 10.4324/9781003500117-17

1. Define different types of reliability
2. Explore Cronbach's alpha and omega reliabilities
3. Demonstrate how to run alpha and omega in SPSS

Different Types of Reliability

As discussed in Chapter 2, in social scientific research, we often use survey-type measures to conduct our research. When using survey-type measures, we must ensure our assessments reliably measure the constructs we are studying. Look back at the example in Chapter 2 about weight and what makes a "reliable" measure. What makes a measure "reliable" is the extent to which it produces consistent and stable measurements. Essentially, if participants take the same measure under the same conditions repeatedly, the results should not be significantly different (reliable). There are different ways we can assess (types) the reliability of scales/measurements.

Test-Retest

The **test-retest** method of reliability is when you give the same measure(s) to participants at different points in time to see if you get the same score. This procedure assesses the stability and similarity of the measure(s) at different times. If you are measuring something that you believe does not change over time (such as a trait-like behavior), then scores on a measurement of that behavior should not change significantly over time. At testing point 1, you would give the measure(s) to the participants and then give the same measure(s) at testing point 2. The two scores on measure 1 (testing point 1 vs. testing point 2) are then analyzed using correlation analysis. The correlation analysis (Chapter 8) will show the degree of association between the two testing points. This result is called the test-retest reliability coefficient. A reliable measure will have a reliability coefficient of .70 or above. The higher the coefficient, the more reliable the measure(s). Reliability coefficients below .70 demonstrate that the participants are not responding reliably or consistently to the measure.

Alternate Forms

The alternate forms reliability test is when you use two or more instruments to measure the same construct/variable. For example, instead of giving a participant one measure of willingness to communicate, you can give them two different measures. Give them measure one at point 1 and then measure 2 at point 2. In this case, you are still assessing at two different periods in time, but you are also using two different measures. Then, just like test-retest, you would run a correlation analysis and assess the strength of the correlation. Reliable measures again have a coefficient of .70 or above. Expect to have slightly lower reliabilities with this approach than with the test-retest method, as you are using two different measures.

Split-Half

If you do not have the time or resources to administer a measure over two time periods, you can use the split-half measure. The **split-half** measure is a one-shot approach that computes two scores for each participant. One score comes from half of the measure, and the other score from the other half of the measure. Let's say you are measuring the tendency to approach arguments (Infante & Rancer, 1982). There are ten items that make up this construct. Half of the score would be five items, and the other half of the score would be the other five items. You would compare the score on each half using a correlation analysis. As with the previous tests, a coefficient of .70 or above is a reliable measure, while a coefficient of below .70 is not reliable.

Test Your Understanding 17.1 – What is the minimum coefficient threshold for reliability?
Test Your Understanding 17.2 – What are the differences between split-half, test-retest, and alternate forms reliability?

Cronbach's Alpha and Omega

While test-retest, alternative forms, and split-half are perfectly acceptable reliability tests, they are not something you are likely to use as a social scientist. These tests are simply less practical for social scientists because they often require repeated testing or dividing measures into equal parts, which makes no sense with the short length of measures we use in the social sciences. Cronbach's alpha and McDonald's omega are better suited for the social sciences because they provide estimates of internal consistency amongst measurement items. The other methods are more commonly used in educational assessments, where longer measures make these approaches more feasible and appropriate.

Cronbach's alpha (α) is a method of reliability that builds on the split-half method. It splits the data in two in every possible way and computes a correlation coefficient for every split. A limitation of the split half is that there are several ways in which the split could occur, leading to bias and error. Cronbach's α bypasses this with infinite splits. (This means not only is Cronbach's α more practical for social sciences, it's also more rigorous.) Cronbach's α results of .70 to .80 are considered acceptable, while results above .80 show strong reliability.

A couple of things to consider when interpreting Cronbach's α. First, Cronbach's α assumes a measure is unidimensional, meaning all items measure one construct from one factor. You will read much more about this in Chapters 18–20. A "reliable" Cronbach's α does not provide evidence of unidimensionality for a multidimensional measure. It is common for researchers who use multidimensional measures to run Cronbach's α on each factor; this is problematic, as Cronbach's α does not communicate dimensionality; factor analysis does this (Goodboy & Martin, 2020). Second, the value is largely dependent on the number of measured items. The more items that are present in the measure, the higher α is likely to be, as part of the α equation includes the number of items squared.

Test Your Understanding 17.3 – What are some of the limitations of Cronbach's α?

While there are some issues to consider when interpreting and running Cronbach's α, it has remained the most common form of reliability reported in social sciences, largely because it is the default reliability analysis in SPSS. Researchers have recently called for the use of McDonald's (1970, 1999) **omega** (ω) (Cho & Kim, 2015; Goodboy & Martin, 2020; Hayes & Coutts, 2020). Without getting too much into the mathematics of it all, omega combines confirmatory factor analysis (CFA), error variances, and factor loadings (validity) into its equation, reflecting the true score variance relative to the total observed variance. More simplistically, unlike Cronbach's α, McDonald's ω does not assume all items on a scale contribute equally to the construct, making it a more nuanced measure of internal consistency. Importantly, McDonald's omega also prevents negative correlations between items from artificially inflating reliability estimates, ensuring that only items that assess the construct contribute to the calculation. Cronbach's alpha has no problem with items that are supposed to measure the same variable being negatively correlated with one another, even though that makes no sense. If the items are measuring the same thing, they must be positively correlated! This means that bad items can make α reliability scores inflated, but ω catches these problems. This makes McDonald's ω a more robust method of reliability that incorporates validity indices into its calculation and is less affected by sample size and the number of items on the scale. Below, we will show you how to compute both α and ω in SPSS. We encourage you, if you are progressing in your research, to look at the debate over these two tests of reliability. Then we encourage you to do what you know is better, even if it requires an extra click in SPSS, and use McDonald's ω when calculating measure reliability.

Test Your Understanding 17.4 – Which reliability calculation should a social scientist use and why?

How to Conduct Alpha and Omega in SPSS

Before running the reliability scores, let's first talk about data preparation. Many measures have reverse-coded items. Read more about reverse-coded items in Chapter 3. When computing reliability scores, reverse-coded items must first be adjusted to align with the same direction as other items to prevent calculation errors. You will need to have only items coded in the same direction, as the divergence essentially creates confusion for the equation.

Notably, when running reliability scores in SPSS, you get an option that says, "if item deleted." This is output in SPSS that shows you the change in a reliability score if an item is deleted. It is common for researchers to delete an item from their analysis to "improve" reliability. However, this procedure does not take into consideration the validity of the measure. We encourage researchers to not drop items based on "if item deleted" information from SPSS.

Running alpha and omega in SPSS is relatively easy and quick. What follows is a reliability analysis of a 4-item measure of perception of acculturation. First, as shown in Figure 17.1, open the Reliability Analysis menu by clicking:

1. Analyze
2. Scale
3. Reliability Analysis

These steps open the window for the reliability tests shown in Figure 17.2. Within this window:

4. Highlight the variables that make up your measure and move them over to Items
5. Click Statistics
6. Select statistics you should for Item, Scale, and Scale if item deleted. This information, shown in Figure 17.3, will show you what impact each item is having on your measure's overall reliability
7. Click Continue

When you are back to the Reliability analysis screen, you will see at the bottom left a box named "Model." Here you can select various types of reliability models. The default is Alpha. Click "OK" and SPSS will run alpha. Those five outputs are as follows (17.1–17.5). Re-run all the steps above but change the model to "Omega" and select "OK" to run omega. Those outputs are Tables 17.1 and 17.2.

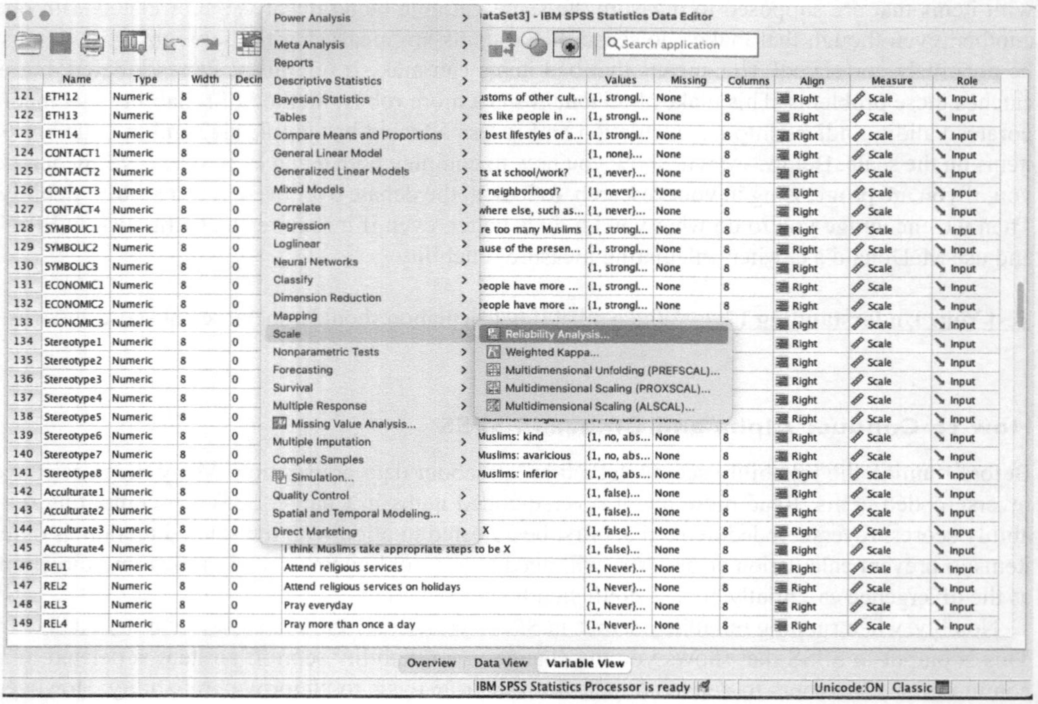

Figure 17.1 Calling up the reliability analysis test in SPSS.

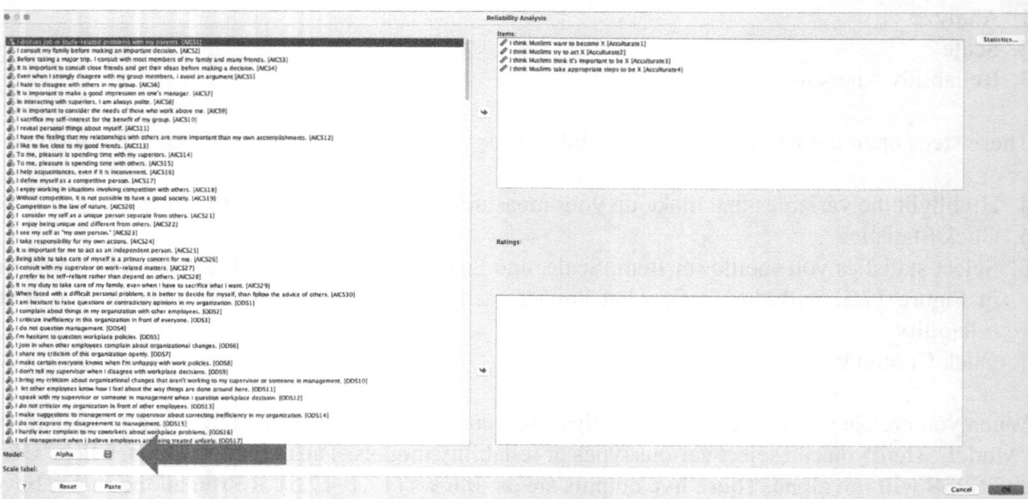

Figure 17.2 Item selection in reliability analysis.

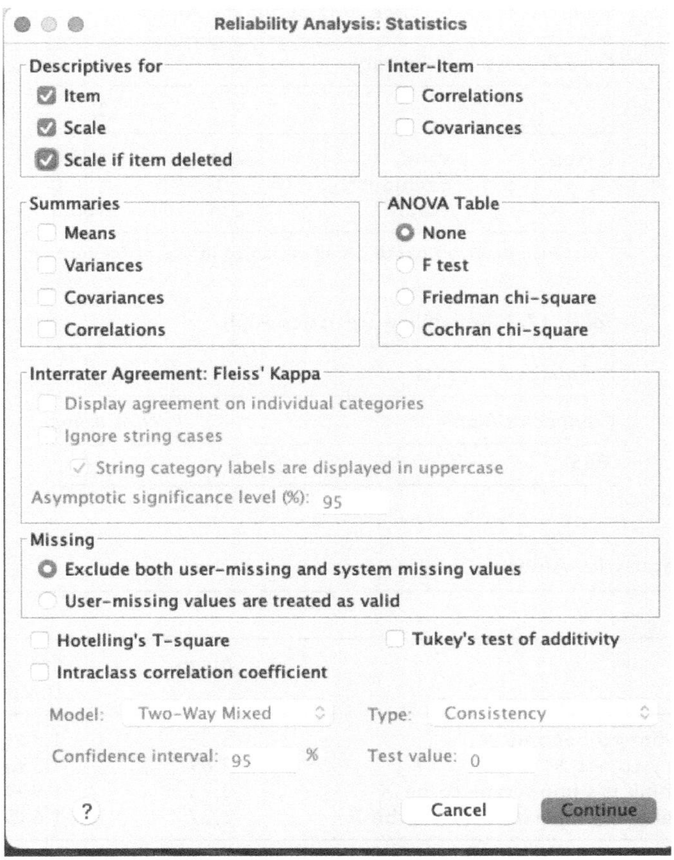

Figure 17.3 Reliability analysis: Statistics.

The first output, Table 17.1, shows the number of cases in the data set.

The second output, Table 17.2, shows Cronbach's alpha for this measure. In this case, the alpha is .96. If you look back to the earlier part of the chapter, that is a very high alpha and thus can be interpreted as a highly reliable measure.

Table 17.3 (Item Statistics) shows you the means, standard deviations, and number of participants who completed each item.

Table 17.4 is the Item-Total Statistics. This output shows the mean and variance of the scale is an item is deleted. In addition, it shows the alpha if an item is removed from the measure.

The final output is the overall scale statistics, essentially the mean, variance, and standard deviation for the measure.

The omega results will be identical for Case Processing Summary, Item Statistics, and Scale Statistics. The differences are below. Table 17.6 is the omega reliability. You will notice that the alpha was .865, rounded up to .87. The omega is .845, rounded up to .85. So, omega is a bit lower.

Second, the Item-Total Statistics (Table 17.7) output for omega reveals a bit more differences in terms of what impact items would have if deleted. As with Table 17.4, the fourth item appears

Table 17.1 Case Processing Summary Alpha

Case Processing Summary

		N	%
Cases	Valid	244	100.0
	Excluded[a]	0	.0
	Total	244	100.0

[a] Listwise deletion based on all variables in the procedure.

Table 17.2 Reliability Statistics Alpha

Reliability Statistics

Cronbach's Alpha	*N of Items*
.865	4

Table 17.3 Item Statistics Alpha

Item Statistics

	Mean	*Std. Deviation*	*N*
I think Muslims want to become X	3.27	1.901	244
I think Muslims try to act X	2.69	1.556	244
I think Muslims think it's important to be X	3.25	1.909	244
I think Muslims take appropriate steps to be X	3.27	1.635	244

Table 17.4 Item-Total Statistics Alpha

Item-Total Statistics

	Scale Mean if Item Deleted	*Scale Variance if Item Deleted*	*Corrected Item-Total Correlation*	*Cronbach's Alpha if Item Deleted*
I think Muslims want to become X	9.20	19.050	.752	.812
I think Muslims try to act X	9.78	22.105	.725	.826
I think Muslims think it's important to be X	9.23	18.636	.780	.799
I think Muslims take appropriate steps to be X	9.20	22.838	.616	.864

Table 17.5 Scale Statistics Alpha

Scale Statistics

Mean	*Variance*	*Std. Deviation*	*N of Items*
12.47	35.139	5.928	4

Table 17.6 Reliability Statistics Omega

Reliability Statistics	
McDonald's Omega	*N of Items*
.846	4

Table 17.7 Item-Total Statistics Omega

Item-Total Statistics						
	Scale Mean if Item Deleted	*Scale Variance if Item Deleted*	*Corrected Item-Total Correlation*	*Squared Multiple Correlation*	*Cronbach's Alpha if Item Deleted*	*McDonald's Omega if Item Deleted*
I think Muslims want to become X	9.20	19.050	.752	.789	.812	.818
I think Muslims try to act X	9.78	22.105	.725	.658	.826	.871
I think Muslims think it's important to be X	9.23	18.636	.780	.806	.799	.805
I think Muslims take appropriate steps to be X	9.20	22.838	.616	.589	.864	.898

to be the one that, if deleted, would improve the measure's reliability the most. However, as stated earlier in the chapter, we do not recommend deleting items based on reliability analysis. We recommend the deletion of items that are linked with validity analysis, which is covered in the following three chapters.

Typically, you would write up the results of a reliability analysis in the method section or in a table form. For example, you could report these results as such: 244 participants completed the 4-item Perception of Acculturation Measure (α =.87; ω =.85). You can also choose to report only omega with the descriptive statistics.

Summary

Looking back, Princess Leia was right. Never underestimate a droid. Throughout the battles between the Republic, the Empire, and the different manifestations of the Empire, droids have been reliable companions and fighters. Leia and the Rebellion were able to rely on R2, C3PO, BB8, and so many others for help and guidance. In this chapter, you've covered the essentials of reliability. How consistent and dependable our measures are over time is critical to our ability to make accurate inferences from our data. In the next three chapters, you will cover another critical issue – that of validity. Validity is essentially asking, "Are we measuring what we say we are measuring?"

References

Cho, E., & Kim, S. (2015). Cronbach's coefficient alpha: Well-known but poorly understood. *Organizational Research Methods*, *18*(2), 207–230. https://doi.org/10.1177/1094428114555994

Goodboy, A. K., & Martin, M. W. (2020). Omega over alpha for reliability estimation of unidimensional communication measures. *Annals of the International Communication Association*, *44*(4), 422–439. https://doi.org/10.1080/23808985.2020.1846135

Hayes, A. F., & Coutts, J. J. (2020). Use omega rather than Cronbach's alpha for estimating reliability: But . . . *Communication Methods and Measures*, *14*(1), 1–24. https://doi.org/10.1080/19312458.2020.1718629
Infante, D. A., & Rancer, A. S. (1982). A conceptualization and measure of argumentativeness. *Journal of Personality Assessment*, *46*(1), 72–80. https://doi.org/10.1207/s15327752jpa4601_13
McDonald, R. P. (1970). The theoretical foundations of principal factor analysis, canonical factor analysis, and alpha factor analysis. *The British Journal of Mathematical and Statistical Psychology*, *23*(1), 1–21. https://doi.org/10.1111/j.2044-8317.1970.tb00432.x
McDonald, R. P. (1999). *Test theory: A unified treatment*. Lawrence Erlbaum.

Chapter 18

Exploratory Factor Analysis

What Will I Learn in This Chapter?

Ever since that unfortunate gamma radiation incident, Lego Dr. Bruce Banner has turned into the Incredible Hulk every time he gets mad. Hulk is strong and unpredictable. Dr. Banner and Hulk are not exactly Jekyll and Hyde, but Dr. Banner certainly prefers to remain himself, totally in control of his behavior. The only way to maintain control is to control his emotions, meaning he must become more aware of what makes him angry and when episodes of anger are likely to occur.

DOI: 10.4324/9781003500117-18

Thanks to the nanotechnology provided by Lego Tony Stark, Dr. Banner has lots of data on what has happened the last 20 times he became angry:

1. Who was present and how far away were they from him
2. His temperature
3. The temperature of his environment
4. His pain index
5. Noise levels
6. Crowd density
7. Blood pressure level
8. Blood sugar level
9. Cortisol level
10. Time since eating
11. Sleep within the last 24 hours
12. Quality of that sleep

At this point, he has no idea which of these variables go together to explain his anger. He never even notices he's starting to get angry . . . he just suddenly starts turning green. Since he has no idea how these variables go together, he looks to an exploratory factor analysis (EFA) to see how these variables cluster together, which of those clusters explain his transition, and how well the clusters explain his transition. Several important clusters of data emerged:

- Stress: Several factors such as blood pressure, cortisol, and crowd density meshed in a factor that worked together to capture how stressed Dr. Banner was directly before these incidents.
- Physical wellness: Other factors such as his quality/quantity of sleep, pain index, and blood sugar levels seemed to contribute to his transition when they were lacking.
- People he likes: Notably, there were different clusters of people who associated with Dr. Banner's transitions. Good friends like Lego Tony Stark and Lego Black Widow seem to be associated with low anger.
- People who get on his nerves: On the other hand, being close to people who get on his nerves seems to contribute to anger.

From this, Dr. Banner learns that managing factors related to stress (like avoiding crowds), taking care of his body, spending time with people he likes, and avoiding people he does not like are overarching categories of variables he should focus on to take care of himself. This means some variables like noise and temperature were not a concern for him, even though it made logical sense that these might be anger triggers.

This chapter will overview:

1. What can be learned from an EFA
2. Assumptions of EFA
3. Best use practices of EFA
4. How to run an EFA in SPSS

What Is Exploratory Factor Analysis?

Factor analysis is a statistical method of identifying underlying relationships between variables. **Exploratory factor analysis** is a type of factor analysis developed for measuring variables that cannot be measured directly (i.e., **latent variables**). In the example from the introduction, you cannot directly observe Dr. Banner's anger. You can see physical cues that his anger is present, but you cannot get directly into his head and use an emotional ruler to measure his anger. As such, anger, like any other emotion, thought, feeling, or perception, is a latent variable.

The purpose of EFA is to describe the relationships among many observed variables (i.e., that which is directly measured) and indirectly observed latent variables (i.e., **factors**). In the example from the introduction, there were several variables that Lego Tony Stark was able to measure for Lego Dr. Banner because they could be directly observed: cortisol level, noise, blood sugar, blood pressure, etc. When Dr. Banner ran the EFA, it produced clusters of these variables that worked together. Those clusters are factors, collections of variables that the researcher names themselves based on what they have in common (e.g., Dr. Banner called the factor composed of blood pressure, cortisol, and crowd density "stress"). When an EFA is conducted, it is because researchers do not know which of the directly observed variables are working in tandem. That is what being part of a factor means – these variables come together in a way the researcher previously did not know.

In EFA, a causal relationship is assumed between the observed and latent variables such that any changes in the observed variables are assumed to have been caused by latent variables. EFA attempts to produce factors that represent true variance, using the calibrations of all considered variables to account for measurement noise versus actual influence. In sum, EFA is the correct analysis for explaining how a latent variable is manifested in a combination of observable variables when the researcher has no basis for understanding how those variables go together (i.e., which variables compose factors).

Test Your Understanding 18.1 – What is the difference between a latent variable and a directly observable variable?

Assumptions

EFA has several assumptions, the first three of which were alluded to earlier in the chapter but will be reiterated for a consolidated list:

1. Linearity: There is a linear relationship among variables.
2. Variance Attribution: Variance in observed variables is due to latent variables and measurement error.
3. Factor Contributions: Each factor contributes to the variance in the explained phenomenon.
4. Adequate Power: Sample size is large enough to achieve a normal distribution ($n \geq 20$).
5. Normality: The factors are normally distributed.
6. Sufficient Correlation: There is some degree of correlation between variables.
7. No multicollinearity: All variables in the model are unique.
8. Independence of Observations: None of the data points should influence another. For example, if you administer a questionnaire to a group of people, the administration should be done silently, without commentary on the items, so no one's opinion of the items influences the way others respond to the questionnaire.

The only assumptions that must be checked prior to running EFA are the assumptions of sample size and normality. The other four assumptions are used in the analysis to create factors.

Rotation Choice

When running an EFA, you will need to choose a rotation method. Rotation methods are like the zoom function on your electronic devices. Using rotations means you are zooming in on factors to get a better look at how closely certain variables are related. EFA rotates data from the factors around coordinate axes until a simple structure is achieved (i.e., observed variables align into a single factor) to determine your output. This should ideally result in a pattern of each observed variable loading onto a single factor highly correlated with the variables that compose it and weakly correlated with every other factor. Unless you have a math degree, that process is likely very difficult to visualize. If that's the case, just think of the rotations as zooming in.

There are two typologies of rotation choices: orthogonal and oblique. Orthogonal rotations assume the factors to be uncorrelated. Oblique rotations allow factors to be correlated. However, an oblique rotation will change to a 90°, orthogonal rotation if, in analysis, it is discovered that the variables are indeed uncorrelated (Osborne, 2015). Note here that "orthogonal" and "oblique" mean the same thing that they did in geometry when you studied angles. Why angle terminology is relevant is because of the work rotating data around the coordinate axes that your computer will be doing in the background.

Deciding between orthogonal versus oblique rotation can be difficult because sometimes it's not clear whether all factors should or should not be correlated. Consider, for example, the Big Five personality traits (openness, neuroticism, agreeableness, conscientiousness, and openness). Openness and extraversion are typically correlated, likely because they are both linked to a body's natural dopamine production triggers (DeYoung & Gray, 2009). Yet, the strength of these correlations differs across the lifespan as dopamine regulation changes with age (Klimstra et al., 2013). Other personality traits like Openness and Conscientiousness lack both theoretical rationale for connection and evidence of consistent correlation. In the case of these personality traits, it is untrue that all variables should be correlated, but also untrue that no variables should be correlated. For Big Five Personality Data, an oblique rotation is the correct choice because if the factors are indeed uncorrelated, then SPSS will change to an orthogonal rotation in its second rotation, producing results very similar to what would be seen if the analysis had begun with an orthogonal rotation. Table 18.1 overviews the options for EFA rotations.

To make your life easy at this point in your statistical education, this book recommends that you use the Direct Oblimin rotation every time without specifying a correlation strength. Direct Oblimin, being oblique, will correct to orthogonal if needed, so you do not have to worry about your confidence in variable relationships before analysis. Further, because of its ease of use in SAS (a programming language for statistics in social sciences that was most used in communication before SPSS was invented), it is the most used rotation method, so the output is easily recognizable and interpreted across social science fields.

Table 18.1 Rotation Options

Orthogonal	*Oblique*
Varimax – Maximizes the variance of squared loadings of factors across variables	Direct Oblimin: Allows the user to specify the strength of correlation among variables, but will adjust to the strength of the correlation as needed if a strength is not specified
Quartermax – Minimizes the number of factors needed to explain each variable	Promax: A simplistic Direct Oblimin without the option to specify correlation strength
Equimax – A compromise between Varimax and Quartermax	

Uses

EFA was originally developed for tasks such as those Lego Dr. Banner was performing in the introduction: take a lot of measures that may or may not be working together and throw them into an EFA to see what comes out factored together. These factors identified in EFA help you understand more holistically how variables you cannot directly observe work. A classic example of EFA as originally designed to operate was Thurstone's work on intelligence conducted in the 1930s. Up to this point, intelligence was primarily considered a single factor . . . as though there was one muscle in the brain fully responsible for how intelligent someone is. Thurstone (1943) found there were several skillsets that make up intelligence, though: listening skills, verbalization skills, quantitative reasoning, spatial reasoning, memory, perceptual speed, and inductive reasoning. As such, Thurstone found being intelligent wasn't a single thing, but a composite factor composed of multiple abilities.

Today, researchers also use EFA for measurement development work. Unfortunately, it's often misused, and that leads to a lot of measurement error in social science fields. The authors of this book are not going to call out any specific examples of measures developed by misusing EFA but will provide you with an example of what not to do here, and you can probably find examples for yourself of works that have used a similar method.

Evil Bad – An Example of Misusing EFA in Measurement Development

Imagine you want to develop a measure of instructor funniness. Instructor humor has a plethora of positive outcomes in the classroom, including student engagement and retention of information (Cundall, 2021). So, to begin your measurement development study, you survey several classes of students and ask them to describe the behaviors of funny professors. From that data, you get back a lot of recurring statements like this:

- My professor tells jokes
- My professor makes extreme facial expressions
- My professor uses memes in the syllabus
- My professor uses memes in PowerPoint presentations
- My professor is very animated
- My professor uses a lot of exaggerated movements
- My professor smiles a lot
- My professor laughs a lot
- My professor uses sarcasm
- My professor uses puns

In measurement development work, particularly in communication, you often see scholars take a large set of repeated behavior descriptions like this, turn those behaviors into survey items, and then run an EFA on those survey items to determine how many items do a good job of capturing the construct and how many dimensions the construct has (based on how many factors are produced in the output).

There is a plethora of problems with this approach to measurement and the use of EFA. Here are the main problems:

1. Unless your sample is diverse enough to represent all types of students, the items produced will only represent the experiences of the type of student you have and the types of instructors those students have been exposed to. This means the dimensions produced may be an artifact of sampling error.

2. How funny a communicator is lies in the perception of the communication receiver. These items describe the behaviors of the message sender but are designed to capture a perception of the receiver. While collecting behavioral descriptions always provides plenty of items for conducting an EFA, items that describe behaviors of a message sender do not equate the perceptions of a message receiver.
3. Collecting data this way assumes everyone has the same definition of funniness. To one person, that could mean someone makes you laugh. To another person, that could mean someone attempts humor. Factors that arise in the data could be an artifact of varied definitions of what it means to be funny, rather than capturing what the researcher is looking for.
4. The primary problem with this data collection method relative to EFA is that if the measurement development team already knows enough about funniness that they need to develop a measure of it, then they already know what funniness is. If the authors already know what funniness is, there should already be a conceptual definition of the term. If there is already a conceptual definition, the authors know how many dimensions the measure should have. (For example, in an earlier chapter, credibility was defined to you as competence, trustworthiness, and caring – a measure composed of three dimensions; McCroskey & Teven, 1999). If this is the case, a definition is already well established, then the measurement development team is not going blindly into the analysis and should already know how many factors to look for, meaning using an EFA to guide the measurement development conforms a definition to the data rather than seeing if the items are of quality to capture the defined variable.

Using EFA to determine how many dimensions a measure has means fitting the definition to the data, which, with this development method, can be affected by sample bias, wording issues, and lack of representation of experiences in the sample. If the conceptual definition of a measure is already known, then the dimensions should be confirmed by the data with a confirmatory factor analysis (Chapter 20) rather than discovered in an exploration (EFA). If in your research practices, you find measures developed this way, scrutinize the items carefully and ask yourself whether these are items that happen to correlate well, or are these items that truly seem to reflect the data definition.

As mentioned earlier, this type of method for measurement development using EFA is common in some social science fields and leads to measurement error. Do not use EFA this way, regardless of how many examples you have seen of this in published work. Scholars learn more about social science measurement development every year. Knowing better means doing better rather than repeating broken methods for the sake of tradition.

How to Use EFA in Measurement Development

You should use EFA in measurement development when you truly do not know how many factors a measure might have. You may find yourself wanting to use a well-established measure in a culture or context it has never been used in before. If the alignment is off, you will have several questions to answer:

- If the measure is used in a new culture, do people in this culture have a different conceptual definition of the variable than people in the culture where the measurement was developed? If so, this measure may not be appropriate at all to use there.
- If used in a new culture or context, does the clustering of items make sense? Perhaps you see one very strong factor and a collection of items that seem lumped together at random. It could be that the items that do not load into the one strong factor have meaning obscured by translation, cultural implications, and/or contextual cues.

- Does the measure have ample power with the items that do load as expected to still capture the construct?

EFA is a way of exploring how well a measure works as intended in a new sample whose cultural norms, language differences, or contextual conditions make it unclear whether the items will still work together the same way.

Likewise, EFA is also a way to check for suspicious behavior in a measure. Many older communication research measures were built with half of the items designed to be reverse-coded (see Chapter 3). A reverse-coded item is one written to capture low agreement with items when the other items are written to capture high agreement. So, for example, in a measure designed to capture teacher funniness, a reverse-coded item may read, "My teacher is not funny." Early communication measurement development work struggled with the wording of reverse-coded items, sometimes accidentally writing items that measured totally different constructs rather than the low presence of the construct of interest. For example, while the item "My teacher is boring" would likely be disagreed with by anyone who thinks their teacher is very funny, whether that professor is boring is not the same as not being funny. They can be very interesting speakers even if they do not use humor. In cases in which reverse-coded items have not been written well, they sometimes accidentally capture a totally separate variable by accident, a **second false factor**.

EFA is a way to detect problems such as second false factors. If you have data that has been collected with a measure in which you are not sure whether the measure captures one or two variables, EFA is the appropriate choice. In this case, you would reverse the score of reverse-coded items in SPSS using the steps shown to change values in Chapter 3. Then you would run the EFA on the data to see if they load into one or two factors. The data should load into one factor if the measure is expected to capture only one factor but will load into two factors if there is a second false factor: one factor of the normally worded items and one factor made of the items designed to be reverse-coded. A classic example of such a measurement issue discovered is Kotowski et al.'s (2009) examination of the Infante and Wigley (1986) verbal aggression measure. The measure was supposed to measure just how aggressive someone's communication was, but the authors found the items designed to be reverse-coded accidentally captured a totally separate variable, verbal benevolence.

Test Your Understanding 18.2 – In what measurement circumstances would it be appropriate to use an EFA?

Interpretation

When your computer is running an EFA, it's using matrix algebra to compute **eigenvalues** from the correlation matrix. Eigenvalues are a measure of the amount of variance in the observed variables accounted for by all the other variables in the same factor. (Well, that's how it's supposed to work. In the case of the reasons why you will likely use EFA, the eigenvalues will tell you how much variance is accounted for by the questionnaire items, rather than variables, that compose individual factors.) In EFA, eigenvalues greater than 1 are considered significant, and are retained in the measurement model.

To look at how to run an EFA, this example utilizes data collected on coworker solidarity. Solidarity is a variable that emerged in interpersonal communication research to describe the simultaneous trust and synchronicity two people may feel toward one another (Wheeless, 1976). Solidarity has been studied in the workplace context before, assessing the way a subordinate feels about their supervisor, and the measure has been shown to not maintain its predicted structure in such assessments. (c.f., Kelly & MacDonald, 2019; MacDonald et al., 2019). There are several reasons the measure may not have worked well when assessing a supervisor–subordinate relationship. First, solidarity takes mutual interpersonal disclosure to be established, and not all supervisors are willing to lower power distances to have these conversations. Second, because solidarity is an assessment of both trust and synchronicity, it could be that, over time, the overlap between those constructs has become blurry in the wording, such that the measure is assessing more than one variable. Third, Wheeless (1976) designed the 20-items measure to have 7 reverse-coded items, and these items could be measuring a second false factor.[1] Last, it is possible some of the items indicate a degree of closeness not possible in workplace settings, such as item 11, which reads, "I love this person." (This last possibility is an example of a context difference that may prevent a measure from working.)

In short, there are a variety of reasons the measure may have been problematic when used to address supervisor–subordinate solidarity. The data examined here assessed solidarity between coworkers. While the second false factor and other validity issues are still possible, so is the contextual issue. However, coworkers would not have the same issue of power distance preventing interpersonal communication. Further, the fact that many full-time employees have a work spouse indicates it is possible to be much closer to a coworker than a supervisor (Whitman & Mandeville, 2021), regardless of the workplace context. In short, there is no way to know how the factor structure would look using the 20-items solidarity measure (Wheeless, 1976) on a set of coworkers. Therefore, an EFA is an appropriate choice to look at such data.

The data analyzed here represents 255 participants who responded to Wheeless' (1976) solidarity measure when thinking about the "coworker you communicate with most often." The measure has 20 items, with items 4, 6, 8, 13, 15, and 18 designed to be reverse-coded. Data were collected on a 7-point Likert-type response scale with responses ranging from *Disagree Strongly* to *Agree Strongly.* Data were analyzed with an EFA using a Direct Oblimin rotation.

The first piece of output is the Kaiser–Meyer–Olkin (KMO) and Barlett's Test table, shown in Table 18.2. The KMO test indicates whether your test has ample statistical power for the EFA results to be trusted. In general, you can move forward interpreting the data if the KMO is .5 or higher. A KMO of .90 or higher, as seen in Table 18.2, is excellent. Bartlett's Test of Sphericity tests the sixth assumption of EFA, that there is some degree of correlation among each variable. A statistically significant Bartlett's Test indicates the variables (or items) are related enough for an EFA to be a meaningful test. The Sig row in Table 18.2 shows the *p*-value associated with the Bartlett's Test, which is less than .05. As such, both the KMO and Barlett's Test indicate you can have confidence in the analysis.

Next, Table 18.3 presents the total variance explained. There are two items to pay attention to here. First is the Total column under the main header Initial Eigenvalues. Values displayed in the Total column show the eigenvalues for each factor. Recall that any eigenvalue greater than 1 is considered significant and should be included in your final factor composition. As such, this data had two factors. Note also the Cumulative % column. This column tells you those two factors together explain 55.8% of the variance in coworker solidarity. Of course, SPSS does not know that

Table 18.2 KMO and Bartlett's Test

Kaiser–Meyer–Olkin Measure of Sampling Adequacy.		0.921394
Bartlett's Test of Sphericity	Approx. Chi-Square	2967.27
	df	190
	Sig.	0

Table 18.3 Total Variance Explained

Factor	*Initial Eigenvalues*			*Extraction Sums of Squared Loadings*			*Rotation Sums of Squared Loadings*
	Total	*% of Variance*	*Cumulative %*	*Total*	*% of Variance*	*Cumulative %*	*Total*
1	8.606	43.029	43.029	8.134	40.670	40.670	7.430
2	2.556	12.780	55.809	2.088	10.440	51.110	5.199
3	0.973	4.867	60.676				
4	0.928	4.641	65.316				
5	0.875	4.374	69.690				
6	0.707	3.534	73.224				
7	0.646	3.230	76.454				
8	0.632	3.158	79.612				
9	0.532	2.658	82.270				
10	0.499	2.493	84.763				
11	0.468	2.341	87.104				
12	0.410	2.051	89.155				
13	0.389	1.945	91.100				
14	0.360	1.799	92.899				
15	0.305	1.523	94.422				
16	0.281	1.407	95.830				
17	0.241	1.206	97.035				
18	0.219	1.095	98.131				
19	0.198	0.989	99.120				
20	0.176	0.880	100.000				

you're measuring solidarity. It is only telling you that these two factors seem to explain 55.8% of the variance in whatever you're assessing. Note that the first factor explains much more variance than the second factor. From this table, you interpret you should retain two factors in your analysis.

The exact same information regarding how many factors should be retained can also be interpreted from the **scree plot**. Scree plots show the trend in eigenvalues to factor relationships. The rule of thumb is if you interpret the number of factors to be retained based on a scree plot, you keep the number of factors to the left of the elbow (where the line flattens). In Figure 18.1, the scree plot has its elbow on factor 3, meaning you should retain two factors.

Test Your Understanding 18.3 – What do scree plots and eigenvalues tell you regarding EFA?

The next table you will see is the Factor Matrix, shown in Table 18.4. The factor matrix shows the correlations between each item and the factor. As these are correlations, those values closest to 1 and −1 are strongest, while those closest to 0 are weakest. This factor loading matrix is provided to you so you can see how SPSS interpreted your data (i.e., how many factors it thought it had and how strongly those items relate to the factor) before it went through the rotation process and zoomed in on those relationships. Although you do not need to make any interpretations of your data using this table, it is useful to see how apparent factors are before moving onto Table 18.5, the Pattern Matrix.

The pattern matrix shows the same relationships again, but this time adjusted to account for the correlation between factors. Look to the pattern matrix to interpret your data. Here, you are looking for two things. One, you need to find any items that did not load higher than .4 (or lower than −.4)

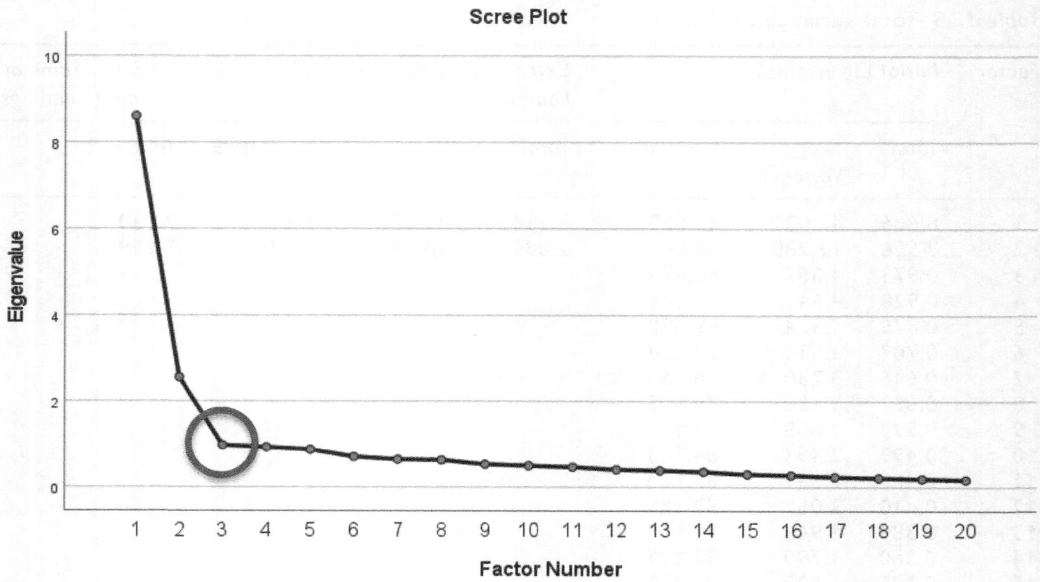

Figure 18.1 Scree plot.

Table 18.4 Factor Matrix

	Factor	
	1	*2*
Solidarity 1	0.822	0.099
Solidarity 2	0.467	0.357
Solidarity 3	0.733	0.139
Solidarity 4	0.380	−0.386
Solidarity 5	0.616	0.270
Solidarity 6	0.596	−0.532
Solidarity 7	0.603	0.274
Solidarity 8	0.692	−0.547
Solidarity 9	0.592	0.322
Solidarity 10	0.430	−0.305
Solidarity 11	0.726	0.244
Solidarity 12	0.677	0.173
Solidarity 13	0.470	−0.547
Solidarity 14	0.651	0.353
Solidarity 15	0.728	−0.298
Solidarity 16	0.717	0.095
Solidarity 17	0.641	0.074
Solidarity 18	0.561	−0.249
Solidarity 19	0.846	0.235
Solidarity 20	0.573	0.347

Table 18.5 Pattern Matrix

	Factor	
	1	*2*
Solidarity 1	**0.683**	−0.258
Solidarity 2	**0.639**	0.160
Solidarity 3	**0.652**	−0.178
Solidarity 4	−0.045	**−0.560**
Solidarity 5	**0.675**	0.006
Solidarity 6	−0.010	**−0.804**
Solidarity 7	**0.669**	0.016
Solidarity 8	0.048	**−0.861**
Solidarity 9	**0.701**	0.070
Solidarity 10	0.060	**−0.499**
Solidarity 11	**0.734**	−0.069
Solidarity 12	**0.638**	−0.120
Solidarity 13	−0.114	**−0.763**
Solidarity 14	**0.770**	0.076
Solidarity 15	0.282	**−0.622**
Solidarity 16	**0.603**	−0.216
Solidarity 17	**0.529**	−0.204
Solidarity 18	0.201	**−0.499**
Solidarity 19	**0.814**	−0.130
Solidarity 20	**0.708**	0.104

in any factor. These items are too weak to meaningfully contribute to your understanding of the factors and should be removed. Second, you are looking for any items that loaded at higher than .4 (or lower than −.4) on more than one factor.

Looking at the output from this EFA, you can see there are no items that loaded weakly on all factors or strongly on more than one factor. If you found this to be the case, then you would have dropped those items, rerunning the EFA without them because those items were simply too problematic in terms of validity to be useful. While no such problematic items were present in this data, a familiar pattern does arise. Notice that items 4, 6, 8, 13, 15, and 18 load as separate factors from factor 1. Those are the items designed to be reverse-coded. From this analysis, you determine that there are at least two separate variables being measured here. Whether those variables are subdimensions of solidarity or a second false factor cannot be determined by the numbers alone. This is where it becomes your job as an analyst to interpret the patterns against what else you know. All SPSS knows is these two factors are separate. The most likely explanation, given the reverse-coded nature of these items, is those items designed to be reverse-coded are, in fact, measuring a second false factor and should not be used to assess solidarity among coworkers.

EFA in SPSS

To run an EFA in SPSS, the first step is to pull up the Factor Analysis options window. This is done with the following steps:

1. Click Analyze
2. Click Dimension Reduction
3. Click Factor

Figure 18.2 shows how to navigate through these steps.

The Factor Analysis option menu is shown in Figure 18.3. The next step in completing the EFA is to repeat the steps used previously to tell SPSS which items will be analyzed:

4. Click on the items you wish to be analyzed in the box to the left
5. Use the arrow beside the Variables box to move items into this box

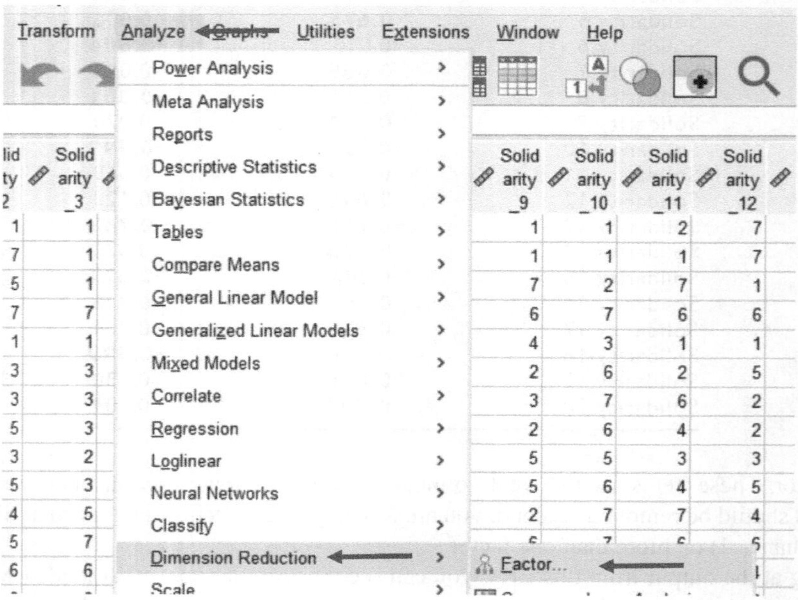

Figure 18.2 Factor menu.

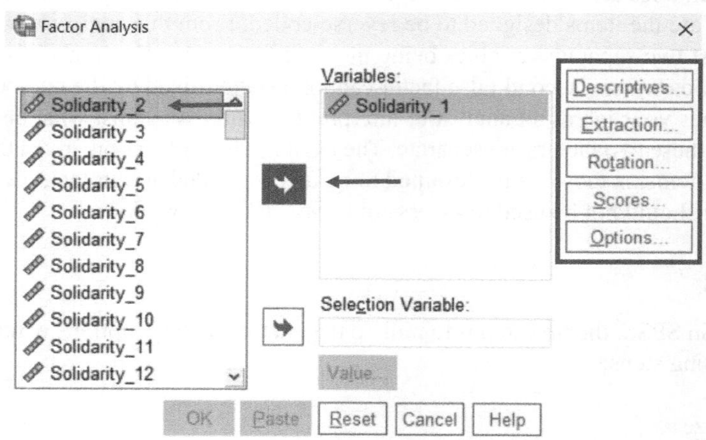

Figure 18.3 Factor analysis options.

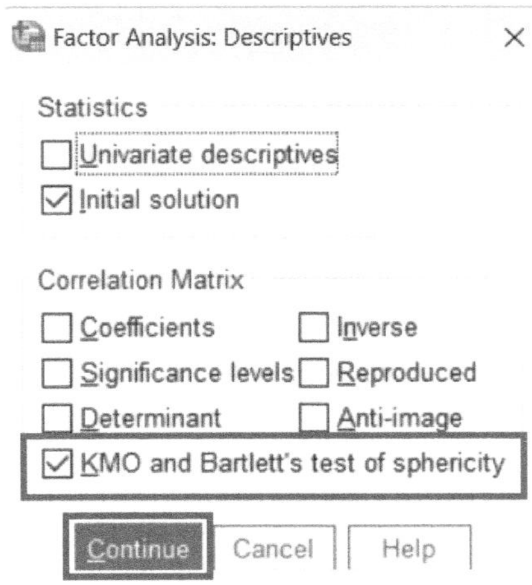

Figure 18.4 Factor analysis: descriptives.

Once your items are selected for analysis, it is time to set up the options for output and model testing. To begin:

6. Click on Descriptives . . . This opens the box shown in Figure 18.4
7. Click the box next to KMO and Barlett's Test of Sphericity to ensure that output is provided
8. Click Continue

The next step is to choose the extraction. "Extraction" is a fancy way of saying choose an algorithm to use. To do this:

9. Click Extraction . . . as shown in Figure 18.3. This will open the Extraction window shown in Figure 18.5
10. Here, you will choose Maximum Likelihood or Generalized Least Squares from the Method dropdown menu. Maximum Likelihood can handle much more complicated data than Least Squares. Generalized Least Squares cannot handle complex, multifactored data well, but can accommodate for your data in cases of non-normality. So, if you have data with some skew or kurtosis, Generalized Least Squares is a better choice. Choose Maximum Likelihood or Generalized Least Squares, depending on which best suits your data
11. Under Display in this window, click the boxes beside Unrotated factor solution and Scree plot to be sure those outputs are provided
12. Under Extract, your model should default to looking for Eigenvalues greater than 1. If by chance this hasn't happened, type "1" into the box next to the Eigenvalues greater than row
13. Also, there should be a number in the Maximum Iterations for Convergence box. This means that your computer stops trying to find a solution if none exists. In the rare case that the data is very strange and has no solution (i.e., patterns are random), your computer will keep analyzing data until it overheats if no number is in this box. The default of 25 usually appears in this box, which is sufficient. Always be sure that there is a number in that box, whether you use the default of 25 or type in a larger number, as a safety measure for your machine
14. Click Continue

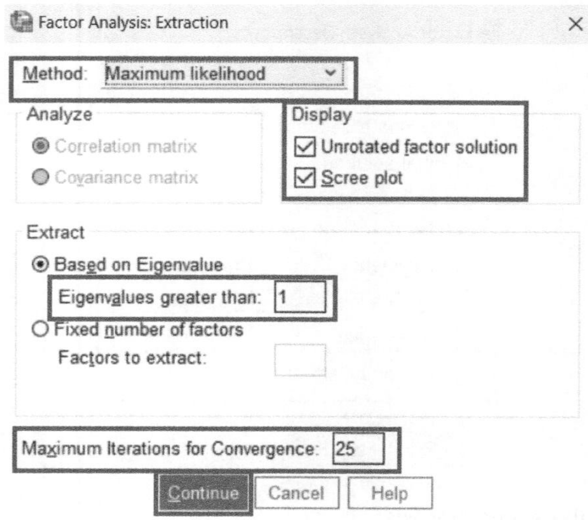

Figure 18.5 Factor analysis: extraction.

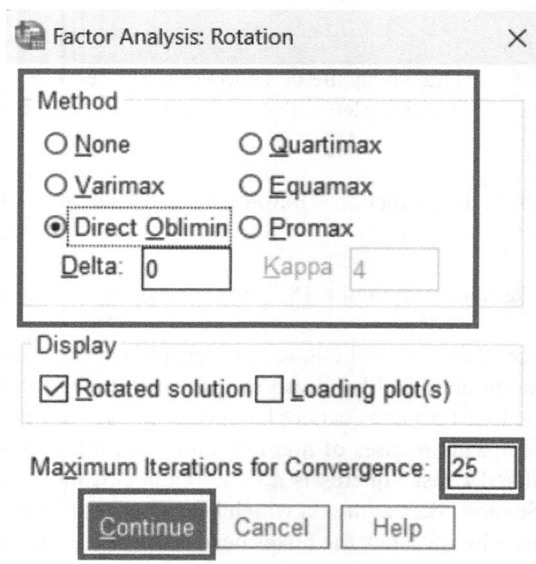

Figure 18.6 Factor analysis: rotation.

The last step will be to select your rotation method. To do this:

15. Click Rotation . . . back in Figure 18.3. This will open the window shown in Figure 18.6
16. Under Method, select Direct Oblimin. Underneath, this should cause a checkmark to appear by Rotated solution under Display
17. Click Continue
18. Click OK

Summary

EFA is the correct analysis to choose when you cannot see the variable you are interested in directly but believe it may be present somehow in some collection of other variables or items. It is the method used when there are no prior assumptions about which variables or items should go together. Like Lego Bruce Banner, who had no idea which of the variables he measured might explain his anger, you can use EFA when you have data but no idea which is important or to what extent. Remember, you've also learned what not to do with EFA, which is to allow it to confirm groupthink in measurement development, creating factors based on the experiences of a few or a misunderstanding of definitions. When scholars learn *how* to do better, it's time to adopt better practices.

Note

1 Notably, in the 1970s, no one had ever heard the term "second false factor," so the fact that Wheeless included these items in the measure should not be thought of as a poor design choice. At the time, all communication scholars believed that including reverse-coded items was a best practice for measurement because more items led to higher reliability scores in an era when the best quality indicator of a measure we had was Cronbach's alpha. If we, as a discipline, are lucky, the calibration of communication measures will continue to improve, and 50 years from now, the measures developed by your authors will be critiqued for their myriad flaws, despite being developed with the best practices of today.

References

Cundall, M. (2021). *Humor, teaching, and digital spaces: Shared backgrounds, emoji's, and the importance of Poe's Law*. Cambridge Scholars Publishing.

DeYoung, C. G., & Gray, J. R. (2009). Personality neuroscience: Explaining individual differences in affect, behaviour and cognition. In *The Cambridge handbook of personality psychology* (pp. 323–346). Cambridge University Press. https://psycnet.apa.org/doi/10.1017/CBO9780511596544.023

Infante, D. A., & Wigley III, C. J. (1986). Verbal aggressiveness: An interpersonal model and measure. *Communications Monographs, 53*(1), 61–69. https://doi.org/10.1080/03637758609376126

Kelly, S., & MacDonald, P. (2019). A look at leadership styles and workplace solidarity communication. *International Journal of Business Communication, 56*(3), 432–448. https://doi.org/10.1177/2329488416664176

Klimstra, T. A., Bleidorn, W., Asendorpf, J. B., van Aken, M. A., & Denissen, J. J. (2013). Correlated change of Big Five personality traits across the lifespan: A search for determinants. *Journal of Research in Personality, 47*(6), 768–777. https://doi.org/10.1016/j.jrp.2013.08.004

Kotowski, M. R., Levine, T. R., Baker, C. R., & Bolt, J. M. (2009). A multitrait–multimethod validity assessment of the verbal aggressiveness and argumentativeness scales. *Communication Monographs, 76*(4), 443–462. https://doi.org/10.1080/03637750903300247

MacDonald, P., Kelly, S., & Christen, S. (2019). A path model of workplace solidarity, satisfaction, burnout, and motivation. *International Journal of Business Communication, 56*(1), 31–49. https://doi.org/10.1177/2329488414525467

McCroskey, J. C., & Teven, J. J. (1999). Goodwill: A reexamination of the construct and its measurement. *Communications Monographs, 66*(1), 90–103. https://doi.org/10.1080/03637759909376464

Osborne, J. W. (2015). What is rotating in exploratory factor analysis? *Practical Assessment, Research, and Evaluation, 20*(1). 2. https://doi.org/10.7275/hb2g-m060

Thurstone, L. L. (1943). *Primary mental abilities*. University of Chicago Press.

Wheeless, L. R. (1976). Self-disclosure and interpersonal solidarity: Measurement, validation, and relationships. *Human Communication Research, 3*(1), 47–61. https://doi.org/10.1111/j.1468-2958.1976.tb00503.x

Whitman, M. V., & Mandeville, A. (2021). Blurring the lines: Exploring the work spouse phenomenon. *Journal of Management Inquiry, 30*(3), 285–299. https://doi.org/10.1177/1056492619882095

Chapter 19

Principal Components Factor Analysis

What Will I Learn in This Chapter?

In the last chapter, Lego Dr. Banner identified multiple factors that contribute to his anger, which causes his transition to the Incredible Lego Hulk:

- Stress
- Physical wellness
- Proximity to people he likes
- Proximity to people he dislikes

DOI: 10.4324/9781003500117-19

Recall that these factors are each composed of multiple variables that cluster together and that the labels of these factors were determined by Dr. Banner after he considered what those groups of variables had in common.

Now, Lego Dr. Banner wants to know which of these factors to focus on managing first. He runs a principal components analysis on the data from those factors retained in EFA. As it turns out, 81% of the variance seems to be explained in a combination of how close he is to people who get on his nerves and his physical well-being. That means the most important things he can do to keep Lego Hulk at bay are to stay on top of his sleep routine and diet (which are the variables in the physical wellness factor) and avoid unnecessary time with Lego Deadpool (who gets on his nerves).

In this chapter you will read:

5. What can be learned from an PCA
6. Assumptions of PCA
7. Best use practices of PCA
8. How to run a PCA in SPSS

Principal Components Analysis

EFA is different from **principal components analysis (PCA)**, though both have been widely used in measurement development. PCA is concerned with explaining the variance-covariance structure of a set of variables with a smaller set of variables. In other words, in PCA, the goal is not to identify which items are measuring what factors, but rather to reduce the number of factors (or items) *needed* to measure the latent variable. Ideally, PCA will identify a subset of observed variables that explains as much variance as the entire set. Also ideally, this new subset of variables is easier to explain (i.e., interpret) than the entire set of variables together. For example, Vaccari et al. (2015) used PCA to explain which political events drove people to engage in dual screening (i.e., being active viewers on traditional media and social media simultaneously). The researchers originally considered 13 types of political activities that could lend themselves to such events, but through PCA discovered collecting data from one of those types of events, using the internet to encourage other people to take political action, did not explain enough unique variance in people's media behaviors for political scholars to study in that context. In short, PCA allowed these researchers to narrow the number of political variables that prompt media consumers to engage in dual screening to a more manageable number of variables to focus on during future data collection.

Some scholars refer to PCA as a descriptive technique rather than a modeling technique like EFA (Unkel & Trendafilov, 2010). This is because PCA **components** (analogous to factors in EFA) are explained by both the true variance and error variance of their observed variables, whereas EFA attempts to produce factors that represent just the true variance. This is why the variance explained by PCA will always be higher than the variance explained by EFA in the same dataset.

Chumney (2012) summarizes the difference between PCA and EFA by saying, "in PCA, extracted components are linear combinations of the raw data, and in EFA extracted factors are an explanation of the raw data and the relationships within them" (p. 7). Both analyses are an attempt to reduce observable data into groups of variables that better explain or measure the unobserved variables. Yet, they have very different purposes. PCA is the correct analysis for variable (or item) reduction. EFA is the correct analysis for explaining how a latent variable is manifested in a combination of observable variables.

Assumptions

Aside from the goals of EFA and PCA differing, these analysis techniques also differ in assumptions. EFA and PCA agree on most data assumptions:

1. Variance Attribution: Variance in observed variables is due to latent variables and measurement error.
2. Factor Contributions: Each factor contributes to the variance in the explained phenomenon.
3. Ample Power: Sample size is large enough to achieve a normal distribution ($n \geq 20$).
4. Sufficient Correlation: There is some degree of correlation between variables.
5. No multicollinearity: All variables in the model are unique.
6. Independence of Observations: None of the data points should influence another.

However, while EFA requires normality of all variables, PCA only prefers it. PCA gives cleaner output with normalized data but can handle considerable skewness and kurtosis if other assumptions are met.

EFA and PCA also assume differences about how what you are observing relates to what you're trying to measure. In EFA, a causal relationship is assumed between the observed and latent variables such that any changes in the observed variables are assumed to have been caused by latent variables. PCA makes the opposite assumption, that changes in the latent variable are caused by changes in the observed variables. So, the underlying assumption of PCA is that measurement models look like what is depicted in Figure 19.1, whereas EFA assumes measurement models appear as depicted in Figure 19.2. Note these models say "Item" to reflect the way communication research most often uses PCA (and EFA), where observed variables are items on a questionnaire. "Variable" in these figures represents the latent variables that EFA attempts to explain and PCA attempts to reduce the items needed to assess.

In terms of what this distinction about measurement assumptions means for you running your analysis, it doesn't mean a great deal. This assumption is explained strictly to give you an idea of how drastically different these types of factor analysis are, and therefore, how important it is to choose the appropriate analysis for your data. The differences between PCA and EFA can become

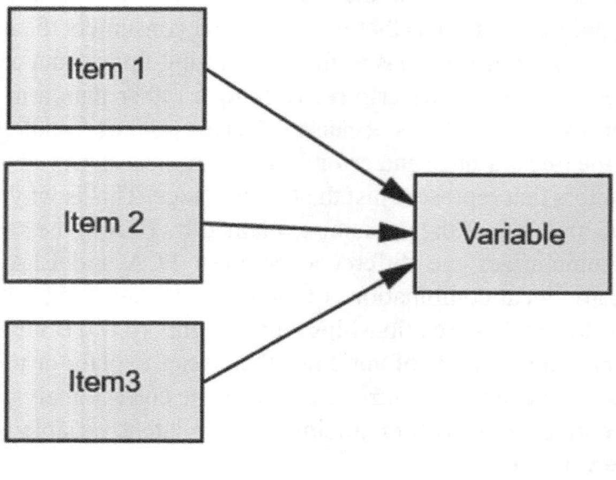

Figure 19.1 PCA measurement model.

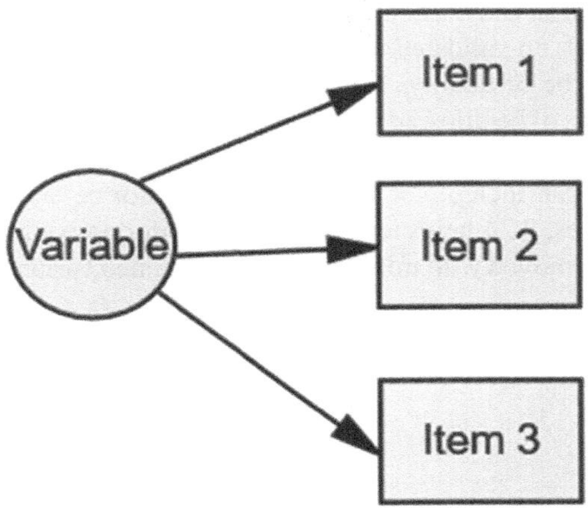

Figure 19.2 **EFA measurement model**

lost given how similar the output looks and how similar the analysis steps are in SPSS, so do not confuse them for the same analysis.

Test Your Understanding 19.1 – How are EFA and PCA different?

Uses

In terms of measurement development, PCA is the appropriate assessment to use if you're already confident your measurement works, but you have a need for it to be shorter. Think, for example, of the standardized exams you've taken throughout your life (e.g., TOEFL, SAT, ACT, EOG). When the first drafts of those tests were designed, there were *far* more items than you saw. Those exam questions were divided into smaller subsets and pilot tested by adding a few extra questions into each approved exam. If you didn't know that you were responding to bonus problems on those standardized exams that didn't count toward your score but were instead being piloted as items on future exams, you can pause here to vent. Eventually, test designers had enough data on each of their items for the new exam so they could look at how the new exam items work overall. Exam designers always want to learn as much as they can about what students know but also make the exam as short as possible for the sake of costs (because the longer those exams must be proctored, the more proctors must be paid, and the less money the exam companies get to keep). So PCA is used for such exams to see how fewer exam questions can be included to fully comprehend what a learner knows.

There are numerous other instances in which measures must be as short as possible. Think about psychological diagnostics at a hospital. Those assessments must be as short as possible because time is of the essence for identifying whether the patient is in danger or could be a danger to others. Likewise, field assessments for the military need to be as short as possible to provide optimal data in minimal time for care decisions.

In terms of practical needs for social science, you're likely to find a PCA to be useful if you do surveys in the workplace. Obviously, being a well-trained student in the social sciences, you will *never* make up your own measure for a questionnaire. If you're supposed to put together an assessment of job satisfaction and burnout, unlike people who haven't been trained better, you will go to Google Scholar, look up validated measures of these variables, and use those, so you know the data you collect will be of quality. However, you might find that those validated measures will take more time than will be allowed for completing the questionnaire. Imagine, for example, you're given five minutes at the end of a company workshop to collect data. That means there are only so many items you can include, or five minutes will not be enough time to complete your questionnaire. In such cases, PCA helps identify the most informative items, allowing you to determine which ones can be removed with minimal loss of explained variance if trimming is necessary.

Test Your Understanding 19.2 – When would you choose to use a PCA?

Interpretations

As mentioned before, running a PCA is very similar to running an EFA. As a factor analytical technique, PCA uses rotations in the same way, following the same logic as EFA, as explained in Chapter 18. This means you can still choose Direct Oblimin each time. Likewise, the interpretation of output is very similar.

For this example, we will use Glaser et al.'s (1987) organizational culture survey. The measure is composed of six sub-measures with 7-point response scales ranging from *Disagree Strongly* to *Agree Strongly*: Teamwork (8 items), Morale (7 items), Information Flow (4 items), Involvement (4 items), Supervision (6 items), and Meetings (5 items). Together, these 34 items explain the culture of an organization. Imagine that you wish to use this assessment at the end of a company retreat, but you know people are going to be exhausted at the end of the day when they receive this assessment, and as such you need to build the questionnaire as concisely as possible. You can use data from a prior administration of the questionnaire to see which items explain most of the variance. For this exercise, a PCA will be run on the data using this measure from a sample of 255 participants.

Output from a PCA will remind you of output from the EFA. To begin, Table 19.1 shows KMO and Barlett's Test. As in Chapter 18, the goal is to see a KMO of .5 or higher to know the data has ample power to trust the output. Likewise, the goal is to have a statistically significant Bartlett's Test of Sphericity to know the components are appropriately related. Both outcomes indicate the data are appropriate for assessing the PCA results.

The next table to observe is again the Total Variance Explained table. As with EFA in Chapter 18, the Total Variance Explained table shown in Table 19.2 displays the eigenvalues for each component and the cumulative percent of variance explained by each component. As expected in this measure of six dimensions, there are six components identified in this measure. In total, all six explain 72.2% of the variance in organizational culture.

Recall from Chapter 18 that a scree plot shows the relationship between the number of things measured (here, components) and eigenvalues. The scree plot tells you how many components or

Table 19.1 KMO and Barlett's Test

Kaiser–Meyer–Olkin Measure of Sampling Adequacy.		0.947
Bartlett's Test of Sphericity	Approx. Chi-Square	7860.195
	df	561
	Sig.	0.000

Table 19.2 Total Variance Explained

Component	*Initial Eigenvalues*			*Extraction Sums of Squared Loadings*			*Rotation Sums of Squared Loadings*
	Total	*% of Variance*	*Cumulative %*	*Total*	*% of Variance*	*Cumulative %*	*Total*
1	17.694	52.042	52.042	17.694	52.042	52.042	12.670
2	2.881	8.472	60.514	2.881	8.472	60.514	12.077
3	1.557	4.581	65.095	1.557	4.581	65.095	9.766
4	1.347	3.961	69.055	1.347	3.961	69.055	8.533
5	1.085	3.191	72.246	1.085	3.191	72.246	7.623
6	1.006	2.959	75.205	1.006	2.959	75.205	9.528
7	0.658	1.935	77.140				
8	0.647	1.903	79.043				
9	0.618	1.817	80.861				
10	0.553	1.626	82.487				
11	0.503	1.480	83.966				
12	0.451	1.327	85.293				
13	0.411	1.210	86.503				
14	0.390	1.147	87.649				
15	0.381	1.120	88.769				
16	0.330	0.971	89.740				
17	0.313	0.919	90.659				
18	0.304	0.893	91.552				
19	0.297	0.873	92.425				
20	0.267	0.784	93.210				
21	0.251	0.738	93.948				
22	0.236	0.694	94.642				
23	0.232	0.681	95.323				
24	0.218	0.641	95.964				
25	0.199	0.586	96.551				
26	0.181	0.533	97.084				
27	0.176	0.518	97.602				
28	0.156	0.459	98.061				
29	0.153	0.450	98.511				
30	0.126	0.372	98.883				
31	0.116	0.342	99.225				
32	0.109	0.320	99.544				
33	0.084	0.248	99.792				
34	0.071	0.208	100.000				

factors to retain by looking one component to the left of the scree plot elbow. Note here that the elbow occurs at component 3. Thus, two components should be kept.

If you're paying close attention at this point, you're probably quite confused. A few paragraphs ago, you were told the data supported there being six components as demonstrated by the eigenvalues, and now you were just told the data clearly supports two components, which can be seen by plotting the eigenvalues in the scree plot. This is a contradiction. This exact contradiction is quite common to see in PCA (and on rare occasions in EFA). Knowing how to handle this contradiction is important.

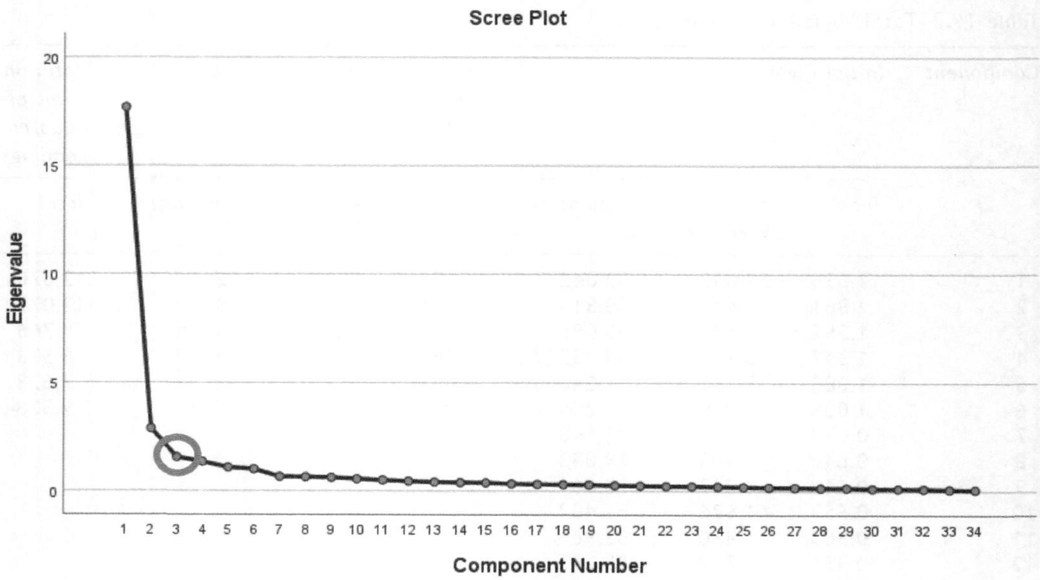

Figure 19.3 Scree plot.

Scree plots and eigenvalue conclusions on how many components to keep will contradict in cases like this, in which there are multiple components with barely significant eigenvalues (multiple eigenvalues under 2). Notice that with these smaller eigenvalues come smaller gains in the variance accounted for. In an ideal world, you would always end PCA accounting for at least 80% of the variance in your latent variable. That is not always feasible in the social sciences, though, where the full measure is sometimes not strong enough to capture that much variance. (This is the case today, but social science measurement gets better every year, so we may be able to expect 80% of variance explained by the next edition of this book.) So, you need to decide on which output to follow. You do this based on determining what is most important to you. Before you analyze your data, decide what your goals and standards are. Possible goals/standards may be:

- You need the most bang for your buck, meaning the fewest items possible that tell you something substantive about your latent variable. In this case, you would go with whichever method suggests you retain the fewest components.
- You need a certain percentage of variance explained to be comfortable with the measure (e.g., 80%) because a high-stakes decision will be made based on this data. In this case, you would make your decision based on the variance explained.
- You are preparing data for an audience without great statistical literacy, and as such, using the scree plot will give them the most concrete understanding of what the data says. In this case, you would make your decision based on the scree plot.
- As scree plots are based on eigenvalues, you would rather trust the original source of data. In this case, you would keep any component with an eigenvalue greater than 1.

In short, your values determine how you interpret the data. Values come before analyses, though. What you do not want to do is look at the data and then decide which standard you want to use to interpret this. Doing this over time will make your analyses inconsistent and hard to defend in a business setting (and even harder in a legal setting). Consistent decisions based on a well-reasoned argument are important for statistical inference.

Regardless of what standard you choose for yourself, when there is disagreement between eigenvalues and the scree plot, you are making a call regarding how much measurement noise you are comfortable with. Imagine that your standard was that *70% of the variance* in organizational culture must be explained. In the case of this data, that means you can lose one component in this data because the first five components explain 72% of the variance in organizational culture. If your standard was you needed to explain *roughly 70% variance*, then in this dataset you could lose two components and still explain 69.1% of the variance. In this example, though, your standard was at least 70% of the variance. That means, given this standard, five components must be retained to explain at least that much variance.

As with EFA, PCA provides a component matrix. This shows how each item correlates with that component. The component matrix shows these correlations before rotation, as shown in Table 19.3.

Table 19.3 Component Matrix

	Component					
	1	*2*	*3*	*4*	*5*	*6*
Teamwork 1	0.690	−0.452	0.069	−0.095	0.076	−0.099
Teamwork 2	0.648	−0.459	0.158	0.073	0.066	−0.077
Teamwork 3	0.703	−0.408	0.040	0.140	0.140	−0.067
Teamwork 4	0.722	−0.433	−0.060	0.156	0.139	0.035
Teamwork 5	0.735	−0.491	0.010	0.067	0.127	−0.007
Teamwork 6	0.720	−0.431	0.051	0.085	0.185	−0.027
Teamwork 7	0.730	−0.376	0.071	0.121	0.266	0.081
Teamwork 8	0.706	−0.489	−0.038	0.126	0.129	0.019
Morale 1	0.736	−0.112	−0.115	−0.049	−0.196	−0.119
Morale 2	0.783	−0.064	−0.136	−0.089	−0.343	−0.116
Morale 3	0.812	−0.089	−0.114	0.019	−0.400	−0.094
Morale 4	0.830	−0.077	−0.079	−0.039	−0.288	−0.041
Morale 5	0.774	−0.136	−0.066	−0.080	−0.241	0.112
Morale 6	0.816	−0.216	−0.087	−0.088	−0.220	0.057
Morale 7	0.777	−0.088	−0.146	0.015	−0.367	0.086
Info Flow 1	0.709	0.168	0.031	−0.090	0.099	0.366
Info Flow 2	0.736	0.150	0.085	−0.171	−0.056	0.276
Info Flow 3	0.626	0.097	0.105	−0.276	0.055	0.554
Info Flow 4	0.756	0.090	0.001	−0.185	0.090	0.346
Involvement 1	0.688	0.147	0.116	−0.364	0.136	−0.057
Involvement 2	0.651	0.137	0.172	−0.405	0.236	−0.322
Involvement 3	0.811	0.247	0.159	−0.283	0.045	−0.202
Involvement 4	0.774	0.173	0.258	−0.337	0.065	−0.230
Supervision 1	0.721	0.269	−0.203	0.226	0.083	0.053
Supervision 2	0.683	0.374	−0.347	−0.019	0.219	−0.066
Supervision 3	0.575	0.346	−0.354	0.241	0.160	0.002
Supervision 4	0.663	0.211	−0.482	0.027	0.080	−0.054
Supervision 5	0.752	0.235	−0.301	0.123	0.042	−0.195
Supervision 6	0.701	0.318	−0.317	0.007	0.213	−0.045
Meeting 1	0.724	0.329	0.095	0.246	−0.043	0.053
Meeting 2	0.725	0.279	0.327	0.185	−0.062	−0.109
Meeting 3	0.587	0.292	0.372	0.484	0.004	0.072
Meeting 4	0.660	0.307	0.409	0.350	−0.036	−0.035
Meeting 5	0.720	0.321	0.377	0.055	−0.114	−0.081

Table 19.4 Pattern Matrix

	Component					
	1	*2*	*3*	*4*	*5*	*6*
Teamwork 1	0.157	−0.701	0.098	−0.085	0.205	0.003
Teamwork 2	0.099	−0.756	0.157	0.097	0.082	−0.051
Teamwork 3	0.032	−0.790	−0.045	0.076	0.022	−0.062
Teamwork 4	0.055	−0.804	−0.113	−0.006	−0.097	0.033
Teamwork 5	0.081	−0.836	0.000	−0.034	0.011	0.029
Teamwork 6	−0.021	−0.829	−0.027	0.026	0.054	0.016
Teamwork 7	−0.163	−0.845	−0.068	0.074	−0.001	0.142
Teamwork 8	0.071	−0.839	−0.051	−0.035	−0.074	0.020
Morale 1	0.550	−0.176	−0.125	0.004	0.115	−0.032
Morale 2	0.769	−0.017	−0.090	0.009	0.109	−0.012
Morale 3	0.828	−0.047	−0.060	0.109	0.001	−0.040
Morale 4	0.672	−0.104	−0.069	0.081	0.063	0.068
Morale 5	0.579	−0.152	0.002	0.005	−0.012	0.248
Morale 6	0.588	−0.258	−0.007	−0.043	0.021	0.185
Morale 7	0.757	−0.038	−0.062	0.055	−0.124	0.160
Info Flow 1	−0.002	−0.074	−0.166	0.122	0.019	0.623
Info Flow 2	0.229	0.032	−0.028	0.131	0.116	0.553
Info Flow 3	0.024	−0.007	0.069	−0.013	0.021	0.905
Info Flow 4	0.062	−0.122	−0.139	0.006	0.084	0.637
Involvement 1	0.031	−0.064	−0.101	−0.002	0.550	0.282
Involvement 2	−0.084	−0.153	−0.129	0.006	0.802	0.010
Involvement 3	0.172	0.015	−0.141	0.184	0.614	0.113
Involvement 4	0.126	−0.054	−0.007	0.181	0.695	0.103
Supervision 1	0.076	−0.085	−0.575	0.246	−0.081	0.104
Supervision 2	−0.026	0.022	−0.780	−0.021	0.195	0.091
Supervision 3	−0.027	−0.014	−0.775	0.137	−0.097	0.014
Supervision 4	0.208	−0.026	−0.758	−0.146	0.021	0.023
Supervision 5	0.231	−0.071	−0.657	0.110	0.115	−0.144
Supervision 6	−0.023	−0.045	−0.729	0.001	0.161	0.101
Meeting 1	0.144	0.025	−0.269	0.538	−0.013	0.130
Meeting 2	0.140	−0.027	−0.036	0.669	0.211	−0.005
Meeting 3	−0.084	−0.119	−0.050	0.880	−0.127	0.048
Meeting 4	0.017	−0.063	−0.003	0.844	0.065	0.006
Meeting 5	0.208	0.098	0.059	0.633	0.303	0.095

The Pattern Matrix (Table 19.4), again like what you saw in Chapter 18, shows the correlations between each item and the latent variable. Here, you look for values with weak relationships (loadings between −.40 and.40) to be removed or values with strong relations (loadings less than −.40 or greater than .40) on more than one component. Like with EFA, these items should be identified and removed from the analysis if present because they show evidence of validity issues.

Note that the data associated with each component are those that load strongly in that component's column. Thus, the components are as follows:

- Component 1: Morale
- Component 2: Teamwork

- Component 3: Supervision
- Component 4: Meeting
- Component 5: Involvement
- Component 6: Information Flow

The ordering of components corresponds to which best explains the variance. This means morale is the most important variable for explaining organizational culture, followed by teamwork, supervision, meeting, involvement, and then information flow. Given the decisions to ensure at least 70% of variance is explained and the results of the Pattern Matrix, the conclusion would be that information flow can be removed from the questionnaire without losing necessary explanatory power.

PCA in SPSS

To run a PCA in SPSS, many of the steps you've already taken will be repeated. To begin, as shown back Figure 18.2:

1. Click Analyze
2. Click Dimension Reduction
3. Click Factor

Then, as shown in Figure 18.3:

4. Click on items that you wish to be analyzed in the box to the left
5. Use the arrow beside the Variables box to move items into this box

Once your items are selected, it will be time to set up the options for output and model testing. To begin:

6. Click on Descriptives. . . This will open the box shown in Figure 18.4
7. Click the box next to KMO and Barlett's Test of Sphericity to ensure that output is provided
8. Click Continue

Next, you will choose your extraction method. As shown in Figure 18.3, Click Extraction . . . Then this will produce the Extraction window shown in Figure 19.4.

9. Once in the Extraction window, select Principal Components as the Method from the drop-down menu. This is where the steps to perform the analysis differ
10. As done in Chapter 18, under Display in this window, click the boxes beside Unrotated factor solution and Scree plot
11. Under Extract, be sure your model is set to 1 next to the Eigenvalues greater than row

Also, as done for EFA and shown in Figure 18.3:

12. Verify that a number is entered in the Maximum Iterations for Convergence box (25 is recommended)
13. Click Continue

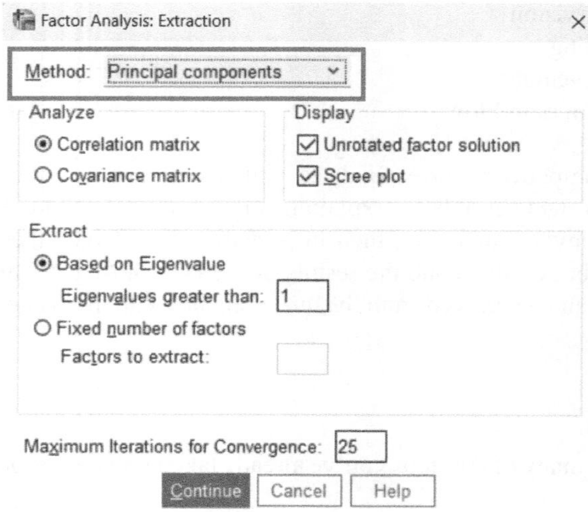

Figure 19.4 Extraction window.

You will also complete the rotation selections as practiced in Chapter 18:

14. Click Rotation. . . as shown in Figure 18.3. This will open the window shown in Figure 18.6
15. Under Method, select Direct Oblimin
16. Click Continue
17. Click OK

Summary

Principal component analysis is a variable reduction technique used to reduce the number of items or variables used at once to explain a latent variable. Like Lego Dr. Banner, sometimes there are too many variables (or items) to focus on at one time, and you triage those that have the biggest impact. For any readers who are graduate teaching assistants, you are encouraged to consider running a PCA on any exams or quizzes you have created. (To set up this analysis, you would need to code the data as 1 for problems students answered correctly, 0 for problems students answered incorrectly, or an appropriate decimal representation for partial credit answers.) You may find you can learn just as much about your students' knowledge with far less grading. You're also encouraged to review resources like Bandalos (2018) to learn more about how to validate exam items.

References

Bandalos, D. L. (2018). *Measurement theory and applications for the social sciences*. The Guilford Press.

Chumney, F. (2012). *Principal components analysis, exploratory factor analysis, and confirmatory factor analysis*. University of West Georgia. www.westga.edu/academics/research/vrc/assets/docs/PCA-EFA-CFA_EssayChumney_09282012.pdf

Glaser, S. R., Zamanou, S., & Hacker, K. (1987). Measuring and interpreting organizational culture. *Management Communication Quarterly*, *1*(2), 173–198. https://doi.org/10.1177/0893318987001002003

Unkel, S., & Trendafilov, N. T. (2010). Simultaneous parameter estimation in exploratory factor analysis: An expository review. *International Statistical Review*, *78*(3), 363–382. https://doi.org/10.1111/j.1751-5823.2010.00120.x

Vaccari, C., Chadwick, A., & O'Loughlin, B. (2015). Dual screening the political: Media events, social media, and citizen engagement. *Journal of Communication*, *65*(6), 1041–1061. https://doi.org/10.1111/jcom.12187

Chapter 20

Confirmatory Factor Analysis

What Will I Learn in This Chapter?

At the Lego School of Witchcraft and Wizardry, Lego Professor Dumbledore has a hypothesis he is desperate to test. When young witches and wizards turn 15, they begin lessons in apparition – disappearing from one place to magically appear in another. Neither students' aptitude in the classroom nor aptitude in athletics has ever been a good predictor of how quickly students are able to learn to apparate. So, Lego Professor Dumbledore predicts that, like so many things in life, a person's ability to apparate is an indirect output of how much anxiety they have about doing the task. After all, there is an abundance of research showing that for many basic life skills (e.g., math, public speaking, broomstick riding), a person cannot perform at their true ability level until they have managed their anxiety about the topic.

DOI: 10.4324/9781003500117-20

Yet, after many hours searching Google Scholar, Lego Professor Dumbledore cannot find a single validated measure of apparating anxiety. He has no choice but to make his own. From his operational definition of apparating anxiety as *the amount of angst one feels when thinking about themselves apparating*, Lego Professor Dumbledore develops ten items that he thinks capture that definition to embed in a questionnaire. He also finds a validated measure of burnout and a validated measure of fear of failure, which he knows are separate variables highly related to apparating anxiety to include in his questionnaire. Then he goes about the work of validating his apparating anxiety measure to see if it works as well as he anticipates. For this work, he will need a confirmatory factor analysis.

What Is Confirmatory Factor Analysis?

Confirmatory factor analysis (CFA) is a statistical technique used to verify that a measure is assessing what it's supposed to assess and only that variable. CFA was originally developed to verify that observable variables used to capture a latent variable all indicated the same construct. For example, CFA has been used in the medical field to verify that various biometric data all reflect the same thing, such as blood pressure, perspiration, heart rate, and cortisol production, all measuring stress.

CFA is one of three types of analyses testing relationships between observed and latent variables, known as **structural equation modeling**. It's quite possible to spend an entire semester on the three types of structural equation models, so this chapter will not fully do CFA justice. Rather, it will give you the basics you need for most social science papers. For further readings on CFA, Kline (2023) is an excellent reference.

In the social sciences, CFA is often used to ensure that a collection of survey items proposed to measure a particular variable assesses that variable and only that variable. In the example from the chapter introduction, Lego Professor Dumbledore has developed ten survey items that he believes will capture apparating anxiety. CFA will test whether all items assess the same construct in the first phase of CFA, the test of **internal consistency**. In the second phase, the test of **parallelism**, CFA will check that the items are only measuring apparating anxiety rather than a construct highly correlated to apparating anxiety. In sum, internal consistency tests ensure that items measure the same construct, and parallelism tests ensure that items measure only that construct.

Test Your Understanding 20.1 – When do you use a CFA rather than a PCA or EFA?

Assumptions

EFA and CFA share many assumptions:

1. Linearity: There is a linear relationship among variables.
2. Factor Contributions: Each factor contributes to the variance in the explained phenomenon.
3. Normality: The factors will be normally distributed.
4. Sufficient Correlation: There is some degree of correlation between variables.
5. No multicollinearity: All variables in the model are unique.
6. Independence of Observations: None of the datapoints or data collection conditions influence one another.

CFA also assumes a latent variable is manifested within the items (or variables) used to represent it. As such, a theoretical CFA model matches a theoretical EFA model, as shown in Figure 19.2.

A difference in CFA assumptions is the sample size required, especially for testing measurement models in surveys. The assumption of power for CFA is that the sample size is large enough to achieve a normal distribution *and* detect small effects. This means that for the type of work a social scientist does, samples of at least 200 participants are needed.

Another assumption is that of **over-identified** power. For a model to be over-identified, there must be more unique correlations than items in a measurement model. The formula for determining how many unique correlations a **unidimensional measurement model** (a measure with one dimension) has is as follows:

$$\frac{n}{n-1}$$

In this formula, n is the number of items a measure has.

If a measure has one item, the formula is solved as

$1(1–1)/2 = 1(0)/2 = 0$

This makes sense because a measure with 1 item cannot have any correlations between items. So, a measure with one item is **under-identified**, meaning that there are more items than correlations between items. If a measure has two items, the formula is solved as:

$2(2–1)/2 = 2(1)/2 = 1$

So, a measure with two items is also under-identified. If a measure has three items, the formula is solved as:

$3(3–1)/2 = 3(2)/2 = 3$

A measure with three items is **just-identified**, meaning there are just as many unique correlations as items. If a measure has four items, the formula is solved as follows:

$4(4–1)/2 = 4(3)/2 = 6$

This means that finally, at four items, the model is over-identified, having more unique correlations than items. This means a shorthand for remembering the rule of assumptions for over-identification is that any unidimensional measurement model should have at least four items. If you have less than four items, you can no longer conduct a CFA.

Test Your Understanding 20.2 – How many items must a unidimensional measure have at a minimum to run a CFA?

Uses

There are primarily two reasons to use CFA:

1. To confirm that a previously validated measure is behaving as expected in a new sample
2. To confirm that a new measure, built according to what is known from theory and prior research, is composed of strong, appropriate items

In short, CFA is used to confirm that a measure is behaving as you expect it to – it confirms whether the factor structure aligns with predictions.

To achieve the first purpose of CFA, reexamining a previously validated measure, the test of internal consistency is needed. Any time a measure of a latent construct is used, there is an inherent hypothesis about the build and composition of that measure (Kelly & Westerman, 2020). As such, even if those hypotheses are not explicitly stated as part of the study, the hypotheses still need to be tested.

To achieve the second purpose of CFA, developing a new measure, both tests of internal consistency and parallelism need to be utilized. These steps can provide multiple evidences of validity for a measure. In the test of internal consistency, showing fit between the hypothesized design of the measure and the fit of the data to that measurement model is evidence of **content validity**. In the test of parallelism, showing that the measure is indeed related to, but separate from, measures of variables theoretically related to it provides evidence of **convergent validity**. Tests of parallelism can be supplemented by adding variables that should have no theoretical relationship to your measure. For example, physical characteristics like age, height, or weight are often (but not always) unrelated to social constructs. Adding these variables that should have no relationship to your measure in the test of parallelism to determine that they are indeed unrelated can show **discriminant validity.**

There are many additional types of validity that can be established for a given measure. As this book is not focused on measurement, there is no space to go into validity evidence in more detail. If you are interested in this topic, Kelly and Westerman (2024) is recommended as an introduction to considering validity and measurement, while Bandalos (2018) is an excellent in-depth resource.

Fit Indices

The fit of a measurement model in CFA is determined by **global fit indices**. A global fit index is one of a variety of statistical tests that determine how well a **hypothesized measurement model** (how theory says the data should fit the measure) fits an **observed measurement model** (how the data fit the measure). The four most used in communication measurement development work are goodness of fit index, comparative fit index, root mean square error approximation, and standard root mean square residual.

Goodness of fit index (GFI) measures the proportion of variance accounted for by the estimated population covariance. In simple terms, this means that the observed model is compared to a theoretically perfect model to see how different the observed model is from one in which all variances were predicted perfectly. It is calculated by comparing the sum of the squared differences between the covariance matrices for the observed and theoretically perfect models. This theoretically perfect model does not exist; it is strictly theoretical. If all variances were predicted perfectly in an observed model, the accuracy would be just as great as a measuring tape for length or a scale for weight. Social science measures are not quite there yet, but GFI estimates how far off in accuracy a measure is.

Comparative fit index (CFI) compares your hypothesized measurement model to the fit of a null model covariance matrix. In other words, CFI compares the fit of your model to a worst-case scenario model where no relationships between variables are estimated. CFI is a bit stronger than GFI because it adjusts for sample size, whereas GFI can sometimes indicate a better fit just because the sample is large. This is because CFI is calculated by comparing the chi-square of the null and hypothesized models, which also accounts for how complex the model is (i.e., how many degrees of freedom it has).

Standard root mean residual (SRMR) uses correlation rather than covariance to test fit. If all items that are supposed to measure the same variable were doing so perfectly, then those items would all be perfectly correlated. (Imagine measuring the width of your device screen with several different tape measures – they should all give you the same length, meaning they are all perfectly correlated.) So, the theoretical correlation matrix for the correlations between items in any measure is a bunch of 1's, whereas in reality, the accuracy of the measure is not that strong. SRMR is calculated by taking the square root of the average squared differences between the observed and theoretical model's correlation matrix.

Root mean square error approximation (RMSEA) assesses how well your data fit the model-implied relationships after adjusting for the complexity of the model. It gives a sense of how much error remains in the model per degree of freedom after estimating the best-fitting parameters. These parameters **include factor loadings**, which represent how strongly each item is associated with its latent variable, **factor variances and covariances**, which represent how much the latent variables vary and relate to each other, and **error variances**, which show how much of each item's variance is not explained by the factor. RMSEA is derived from the chi-square and takes both degrees of freedom and sample size into account. In short, RMSEA is a measure of how different your model is from a perfect fit, adjusted for how complex your model is. It is kind of like testing how different a complicated cake recipe tastes compared to a basic one. RMSEA helps determine whether adding complexity is really improving the end result.

There are a plethora of fit indices that can be considered in CFA, not just the four mentioned here. This chapter covered the four most used fit statistics for social science. For details about additional fit indices, Kline (2023) is an excellent reference.

Test Your Understanding 20.3 – What do global fit statistics tell you? Why do you use more than one?

Interpretation

Once you run a CFA, you will get GFI, CFI, RMSEA, and SRMR scores. In general, GFI should be .9 or higher, CFI should ideally be .95 but is acceptable at .9 or higher, SRMR should be .08 or lower, and RMSEA should be ideally .06 but is acceptable at .1 or lower. If all four measurement parameters are acceptable, the measure is considered to have excellent fit. However, a measure can be used if GFI, CFI, and SRMR are acceptable, but RMSEA alone is off. This is because RMSEA is extremely sensitive to misfit. A measure with fit of only GFI, CFI, and SRMR only has "acceptable" rather than "good" fit. Table 20.1 summarizes the fit standards.

This example will examine the instructor humor orientation measure through CFA. The instructor humor orientation measure was developed in the early 1990s to assess how funny students thought their professor was (Booth-Butterfield & Booth-Butterfield, 1991). Although the measure

Table 20.1 Fit Statistics Heuristics

Fit Statistics	*Excellent*	*Acceptable*
GFI	≥ .90	≥ .90
CFI	≥ .95	≥ .90
RMSEA	≤ .06	≤ .10
SRMR	≤ .08	≤ .08

was developed with excellent practices for the time, measurement development has come a long way in the last few decades, and the technology simply did not exist at the time of this measure's development to run CFAs. So, while Booth-Butterfield and Booth-Butterfield's (1991) conceptual work establishes what the measure should look like, it would not be surprising for CFA to reveal some weak items whose misfit was hidden in weaker available statistical tests of the era.

The humor orientation measure is composed of 17 Likert-type items with a 7-point response scale ranging from *Disagree Strongly* to *Agree Strongly*. The measure is theoretically unidimensional. Figure 20.1 shows the proposed measurement model.

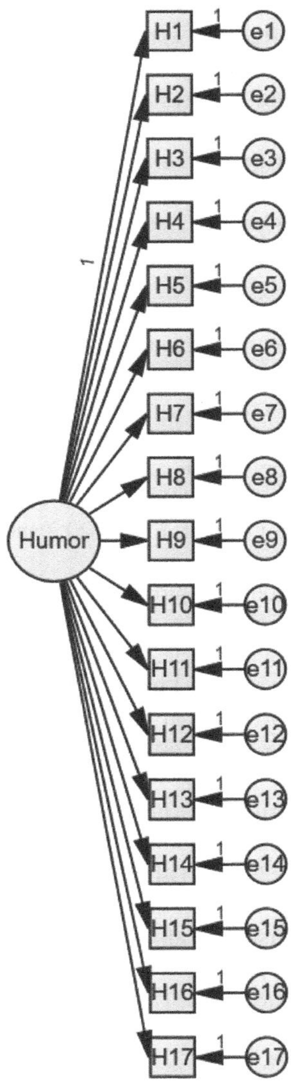

Figure 20.1 Humor orientation proposed measurement model

When the test of internal consistency is conducted, the output is as follows: GFI =.60, CFI =.69, RMSEA =.16, and SRMR =.15. All the fit statistics are poor (i.e., GFI <.9, CFI <.9, RMSEA >.1, and SRMR >.08). This means that the measure must be **respecified**, meaning that problematic items need to be identified and removed. Note here that there is a limit to how much respecification can be conducted on a measure. A measure cannot be respecified, even in a measurement development effort, until it has fewer than four items. If fit is still bad when a measure is respecified down to four items, the measure must be declared inappropriate for use.

Table 20.2 shows the standard residual covariance matrix for this CFA. Now pause! Think back, way back, to Chapter 5. Do you remember that you were told that later in the book, you were going to be expected to remember a few things about *z*-scores and apply them? This is it! This is your moment to use *z*-scores. Now, recall that *z*-scores are a way of standardizing scores on any range of values to the normal distribution. Recall also that *z*-scores smaller than −1.96 or larger than 1.96 are considered abnormally sized (statistically significantly large or small). The standard residual covariance matrix shows you a set of *z*-scores showing the differences in covariances predicted versus observed. That means any values in Table 20.2 −1.96 or above 1.96 represent a statistically significant amount of misfit (i.e., **residual error**).

Test Your Understanding 20.4 – What is a *z*-score? How are *z*-scores useful when conducting a CFA?

Your job interpreting this table is to determine which variable is causing statistically significant residual error with the most other items by counting which variable has the most concerning high/low errors. Both H6 and H3 are causing a statistically significant amount of residual error on nine other items. (Remember that when reading a matrix, items have a row and column associated with them. To help you read this matrix, H6 items have been italicized and H3 items have been bolded.) In cases like this, where items tie in count of problems, you target the item that has the largest residual errors. In this case, focus on H6 and remove it from the model. Removing an item from the model means to literally do just that. Figure 20.2 shows the measurement model less the problematic item.

Now that the most problematic item is removed, fit statistics are reexamined. Fit statistics for the respecified model are as follows: GFI =.66, CFI =.74, RMSEA =.15, and SRMR =.13. The fit statistics are improved, but far from acceptable. Thus, the residual covariance matrix is examined again to decide which item is causing the most trouble. Note that you never remove more than one item per respecification. The standard residual covariance matrix must be calculated anew after each item removal for accuracy. Table 20.3 shows the standard residual covariance matrix after the first respecification, removing item 6.

Notice that if the first standard residual covariance matrix had been used to decide which item would be removed next, the item to cut would have obviously been H3. Now, though, the updated matrix shows that items H3, H10, H13, and H14 each have six problematic items. The magnitudes of *z*-scores in H10 are larger than any others, so this is the item that must be removed in respecification. Fit between items should be considered afresh with each respecification because the relationships between measures change once problematic outliers are removed. Once H10 is removed, the model has better fit: GFI =.74, CFI =.82, RMSEA =.14, and SRMR =.11. Again, these fit statistics are improved, but not acceptable. Table 20.4 shows the standard residual covariance matrix once item 10 is removed.

Table 20.2 Standard Residual Covariance Matrix, First Iteration

	H17	*H16*	*H15*	*H14*	*H13*	*H12*	*H11*	*H10*	*H9*	*H8*	*H7*	*H6*	*H5*	*H4*	*H3*	*H2*	*H1*
H17	0.00																
H16	1.31	0.00															
H15	0.24	0.76	0.00														
H14	−0.48	−4.79	−0.86	0.00													
H13	−0.36	−2.08	−0.90	4.91	0.00												
H12	0.92	3.70	1.17	−3.41	−2.30	0.00											
H11	−0.20	−0.54	1.05	−0.06	−1.10	0.82	0.00										
H10	−1.57	−2.93	−0.26	7.56	6.00	−3.32	−1.71	0.00									
H9	−0.39	−1.94	−0.89	6.45	6.73	−2.40	−1.08	8.01	0.00								
H8	−0.42	−1.69	0.19	1.70	−0.29	−2.13	0.11	1.15	0.72	0.00							
H7	−0.33	0.95	1.05	−3.75	−2.55	1.87	2.68	−3.94	−3.15	−0.67	0.00						
H6	*−0.81*	*−2.03*	*−0.27*	*4.52*	*4.78*	*−4.20*	*−2.21*	*7.28*	*5.49*	*1.80*	*−5.07*	0.00					
H5	0.02	0.05	−0.20	−0.94	−0.54	0.29	0.06	−0.54	−1.06	−0.71	0.25	−0.70	0.00				
H4	−0.09	−0.49	0.25	0.11	−0.46	−0.58	0.52	−0.29	−0.70	0.39	0.81	0.13	0.28	0.00			
H3	**−0.12**	**−3.06**	**−0.73**	**4.65**	**3.11**	**−5.08**	**0.03**	4.53	**3.66**	**2.06**	**−2.37**	***5.19***	**−0.49**	**−0.15**	**0.00**		
H2	−0.66	−0.84	−0.01	0.70	−0.64	−0.85	−0.07	0.83	0.16	0.58	−0.13	1.28	−0.23	0.51	**1.36**	0.00	
H1	0.17	−0.71	−0.31	0.42	−0.69	−1.15	0.23	−0.39	−0.53	0.20	0.25	0.33	−0.03	−0.21	**0.35**	1.24	0.00

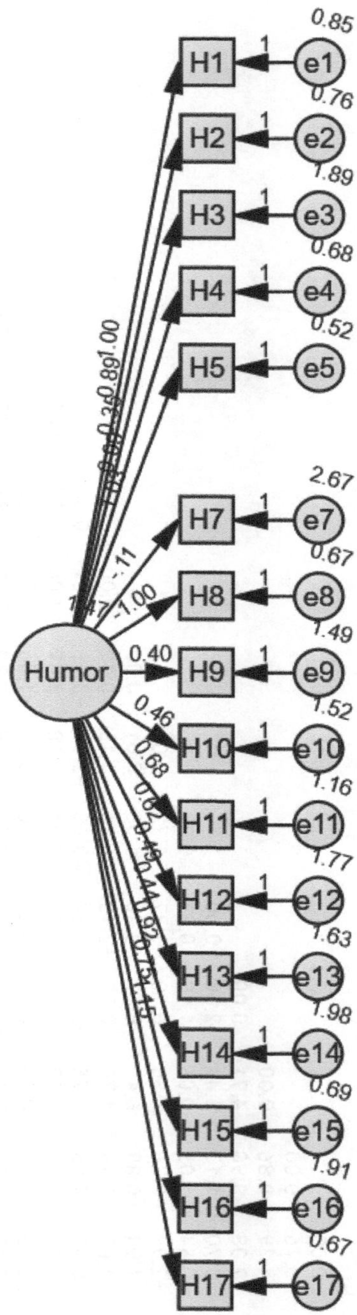

Figure 20.2 Respecified model.

Table 20.3 Standard Residual Covariance Matrix, Iteration 2

	H17	*H16*	*H15*	*H14*	*H13*	*H12*	*H11*	*H10*	*H9*	*H8*	*H7*	*H5*	*H4*	*H3*	*H2*	*H1*
H17	0															
H16	1.17	0.00														
H15	0.18	0.65	0.00													
H14	−0.23	−4.71	−0.67	0.00												
H13	−0.20	−2.01	−0.73	5.12	0.00											
H12	0.70	3.50	0.99	−3.37	−2.28	0.00										
H11	−0.30	−0.67	0.97	0.06	−1.01	0.64	0.00									
H10	−1.31	−2.80	0.01	7.81	6.27	−3.25	−1.54	0.00								
H9	−0.18	−1.84	−0.69	6.66	6.95	−2.35	−0.95	8.29	0.00							
H8	−0.30	−1.52	0.28	1.54	−0.42	−1.89	0.25	0.92	0.54	0.00						
H7	−0.52	0.84	0.88	−3.86	−2.67	1.78	2.56	−4.07	−3.26	−0.50	0.00					
H5	−0.06	−0.08	−0.25	−0.75	−0.37	0.09	−0.04	−0.27	−0.85	−0.58	0.06	0.00				
H4	−0.12	−0.58	0.25	0.32	−0.27	−0.74	0.46	0.00	−0.48	0.46	0.62	0.25	0.00			
H3	0.08	−2.96	−0.53	4.83	3.30	−5.02	0.15	4.76	3.85	1.89	−2.46	−0.28	0.07	0.00		
H2	−0.64	−0.89	0.04	0.94	−0.43	−0.97	−0.08	1.14	0.41	0.60	−0.31	−0.19	0.59	1.60	0.00	
H1	0.15	−0.79	−0.30	0.63	−0.50	−1.30	0.18	−0.10	−0.30	0.26	0.06	−0.04	−0.17	0.57	1.33	0

Table 20.4 shows that the two items with the most problematic *z*-scores are H3 and H14, which both have 6 problematic scores associated with them. The scores on H14 are overall higher. Thus, item H14 is removed to respecify the model. With item H14 removed, fit is again improved: GFI =.80, CFI =.85, RMSEA =.12, and SRMR =.09. Yet, the fit is not strong enough to accept the model. So, a new standard residual covariance matrix is rendered and shared in Table 20.5 less H14.

As seen in Table 20.5, H3 has the most problematic misfit. Thus, H3 must be removed to respecify the measurement model. The fit statistics are again improved: GFI =.84, CFI =.88, RMSEA =.12, and SRMR =.08. For the first time, a single fit statistic, SRMR, is finally within the range of acceptability. Yet, there is still much error in the measurement model, so a new standard residual covariance matrix is computed without H3 and shown in Table 20.6.

As shown in Table 20.6, items H12, H9, and H7 each have three items yielding statistically significant amounts of residual error. As such, one must be removed. Because item H9 has the largest

Table 20.4 Standard Residual Covariance Matrix, Iteration 3

	H17	*H16*	*H15*	*H14*	*H13*	*H12*	*H11*	*H9*	*H8*	*H7*	*H5*	*H4*	*H3*	*H2*	*H1*
H17	0.00														
H16	1.04	0.00													
H15	0.14	0.56	0.00												
H14	−0.08	−4.61	−0.46	0.00											
H13	−0.05	−1.94	−0.57	5.31	0.00										
H12	0.57	3.37	0.89	−3.28	−2.23	0.00									
H11	−0.37	−0.76	0.94	0.21	−0.90	0.54	0.00								
H9	0.02	−1.74	−0.47	6.86	7.16	−2.26	−0.80	0.00							
H8	−0.22	−1.40	0.32	1.35	−0.57	−1.76	0.31	0.34	0.00						
H7	−0.63	0.77	0.77	−3.93	−2.74	1.73	2.48	−3.33	−0.39	0.00					
H5	−0.11	−0.18	−0.26	−0.53	−0.20	−0.03	−0.09	−0.63	−0.54	−0.06	0.00				
H4	−0.15	−0.67	0.26	0.54	−0.10	−0.83	0.43	−0.26	0.48	0.51	0.25	0.00			
H3	0.21	−2.90	−0.39	4.98	3.44	−4.97	0.25	4.00	1.77	−2.51	−0.14	0.22	0.00		
H2	−0.63	−0.94	0.08	1.17	−0.25	−1.04	−0.08	0.64	0.59	−0.42	−0.16	0.64	1.75	0.00	
H1	0.12	−0.87	−0.29	0.84	−0.34	−1.39	0.15	−0.09	0.29	−0.04	−0.04	−0.16	0.71	1.38	0.00

Table 20.5 Standard Residual Covariance Matrix, Iteration 4

	H17	*H16*	*H15*	*H13*	*H12*	*H11*	*H9*	*H8*	*H7*	*H5*	*H4*	*H3*	*H2*	*H1*
H17	0													
H16	0.931	0												
H15	0.127	0.455	0											
H13	0.057	−1.921	−0.473	0										
H12	0.482	3.241	0.802	−2.209	0									
H11	−0.369	−0.829	0.934	−0.819	0.481	0								
H9	0.165	−1.686	−0.345	7.28	−2.214	−0.697	0							
H8	−0.178	−1.269	0.359	−0.66	−1.648	0.328	0.22	0						
H7	−0.719	0.724	0.69	−2.788	1.684	2.418	−3.375	−0.311	0					
H5	−0.129	−0.292	−0.281	−0.101	−0.116	−0.091	−0.494	−0.487	−0.141	0				
H4	−0.138	−0.747	0.267	0.018	−0.9	0.446	−0.112	0.502	0.43	0.255	0			
H3	0.33	−2.864	−0.288	3.537	−4.934	0.333	4.099	1.669	−2.55	−0.028	0.337	0		
H2	−0.602	−1.007	0.103	−0.132	−1.09	−0.048	0.785	0.588	−0.499	−0.137	0.693	1.872	0	
H1	0.14	−0.947	−0.276	−0.226	−1.451	0.173	0.057	0.295	−0.123	−0.033	−0.12	0.829	1.433	0

Table 20.6 Standard Residual Covariance Matrix, Iteration 5

	H17	*H16*	*H15*	*H13*	*H12*	*H11*	*H9*	*H8*	*H7*	*H5*	*H4*	*H2*	*H1*
H17	0.00												
H16	0.88	0.00											
H15	0.13	0.40	0.00										
H13	0.12	−1.91	−0.42	0.00									
H12	0.40	3.15	0.72	−2.22	0.00								
H11	−0.36	−0.86	0.94	−0.78	0.42	0.00							
H9	0.24	−1.66	−0.28	7.34	−2.21	−0.65	0.00						
H8	−0.15	−1.19	0.39	−0.70	−1.54	0.35	0.17	0.00					
H7	−0.76	0.70	0.65	−2.81	1.66	2.39	−3.40	−0.27	0.00				
H5	−0.13	−0.35	−0.29	−0.05	−0.20	−0.09	−0.43	−0.45	−0.19	0.00			
H4	−0.13	−0.79	0.27	0.08	−0.97	0.46	−0.04	0.52	0.39	0.26	0.00		
H2	−0.56	−1.03	0.13	−0.06	−1.14	−0.02	0.86	0.58	−0.54	−0.10	0.74	0.00	
H1	0.16	−0.98	−0.26	−0.17	−1.51	0.19	0.13	0.31	−0.17	−0.02	−0.09	1.49	0.00

Table 20.7 Standard Residual Covariance Matrix, Iteration 6

	H17	*H16*	*H15*	*H13*	*H12*	*H11*	*H8*	*H7*	*H5*	*H4*	*H2*	*H1*
H17	0											
H16	0.849	0										
H15	0.126	0.365	0									
H13	0.26	−1.832	−0.295	0								
H12	0.356	3.087	0.67	−2.153	0							
H11	−0.369	−0.897	0.919	−0.682	0.38	0						
H8	−0.148	−1.153	0.404	−0.833	−1.49	0.365	0					
H7	−0.835	0.655	0.584	−2.855	1.626	2.338	−0.203	0				
H5	−0.132	−0.387	−0.302	0.09	−0.254	−0.109	−0.436	−0.255	0			
H4	−0.119	−0.82	0.267	0.212	−1.015	0.447	0.528	0.32	0.254	0		
H2	−0.54	−1.045	0.145	0.073	−1.17	−0.014	0.57	−0.606	−0.094	0.757	0	
H1	0.174	−1.007	−0.262	−0.032	−1.552	0.185	0.306	−0.231	−0.02	−0.085	1.511	0

z-scores, it is removed. Fit statistics after the removal of item 9 are as follows: GFI =.88, CFI =.93, RMSEA =.10, and SRMR =.06. Now RMSEA has joined SRMR in acceptable ranges. Yet, the measure must be respecified again because of the misfit detected in GFI and CFI. Table 20.7 shows the new standard residual covariance matrix with the removal of H9.

As shown in Table 20.7, both items H7 and H12 are causing statistically significant residual error. Because H12 has the highest *z*-scores, it will be removed. The fit for the measurement model respecified by the removal of item H12 is: GFI =.92, CFI =.95, RMSEA =.09, and SRMR =.05. At last, the measure seems to fit in terms of the test of internal consistency! This means that based on all available data, while there is still measurement noise, the remaining items all seem to be measuring the same thing. The supported measurement model is shown in Figure 20.3.

Notice that in Figure 20.3 there are numbers on each of the lines going from the latent variable to each item. These are the factor loadings! The closer each factor loading is to 1 (or −1), the stronger the item at measuring humor orientation. You may be wondering why there are some positive and some negative numbers. The reason is that the data for this measure has several items that must be reverse-coded before using it for hypothesis testing. CFA can be conducted without reverse coding taking place because it calculates based on magnitude of item relationships rather

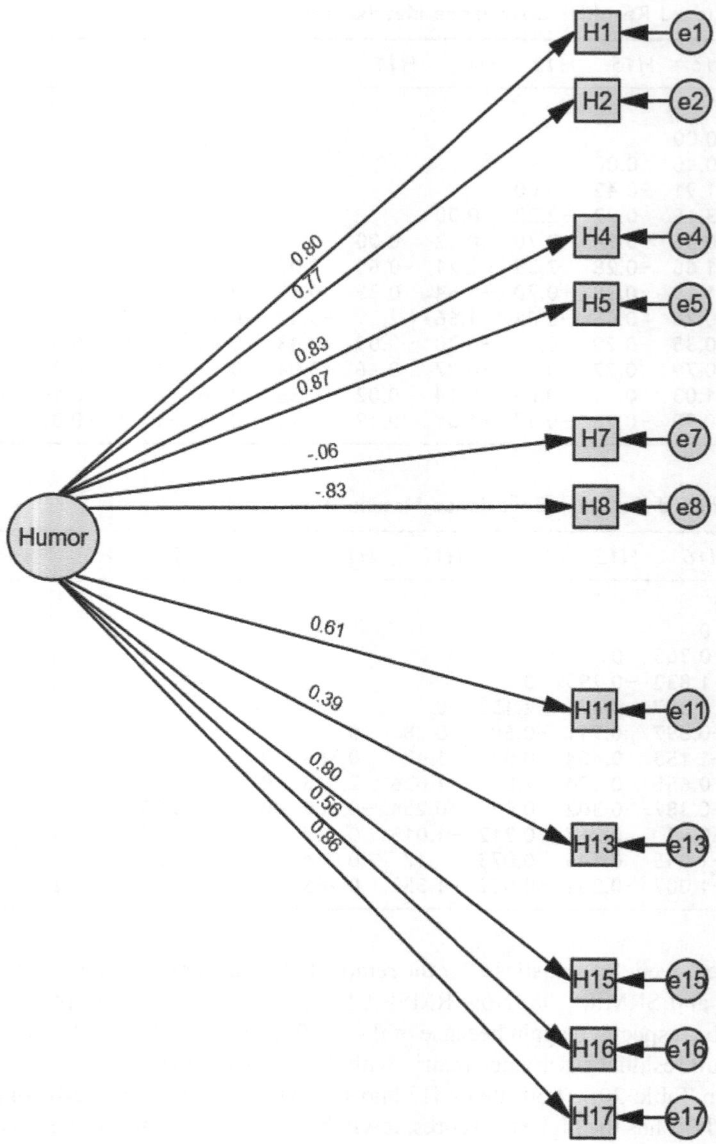

Figure 20.3 Supported measurement model.

than polarity. H7 and H8 would need to be reverse-coded in SPSS before any hypothesis testing took place, though. Also, note that H7, the item that was most problematic of the retained items, has the lowest factor loading. The factor loading indicates what you already knew from the standard residual covariance matrix: that this item is not very good.

Because this measure has not been subjected to CFA before, it is also appropriate to analyze it through a test of parallelism. Perceived immediacy (perceived closeness) and affect for the course content are both likely to be positively related to an instructor's humor orientation (cf. Bolkan & Goodboy, 2015; Violanti et al., 2018). Thus, these measures will be used in a test of parallelism, where we make sure that the items in the humor orientation measure are measuring just humor orientation, and not a construct positively correlated with humor orientation. The perceived immediacy measure is a 9-item semantic differential measure with a 7-point response range (Kelly et al., 2015), and the affective learning toward course material measure is a 4-item semantic differential measure with a 7-point response range (McCroskey, 1994). Both the perceived immediacy and the affective learning measures have been examined through CFA on multiple occasions. In a test of parallelism, each measure is built in the same environment and set to be correlated to one another, as shown in Figure 20.4.

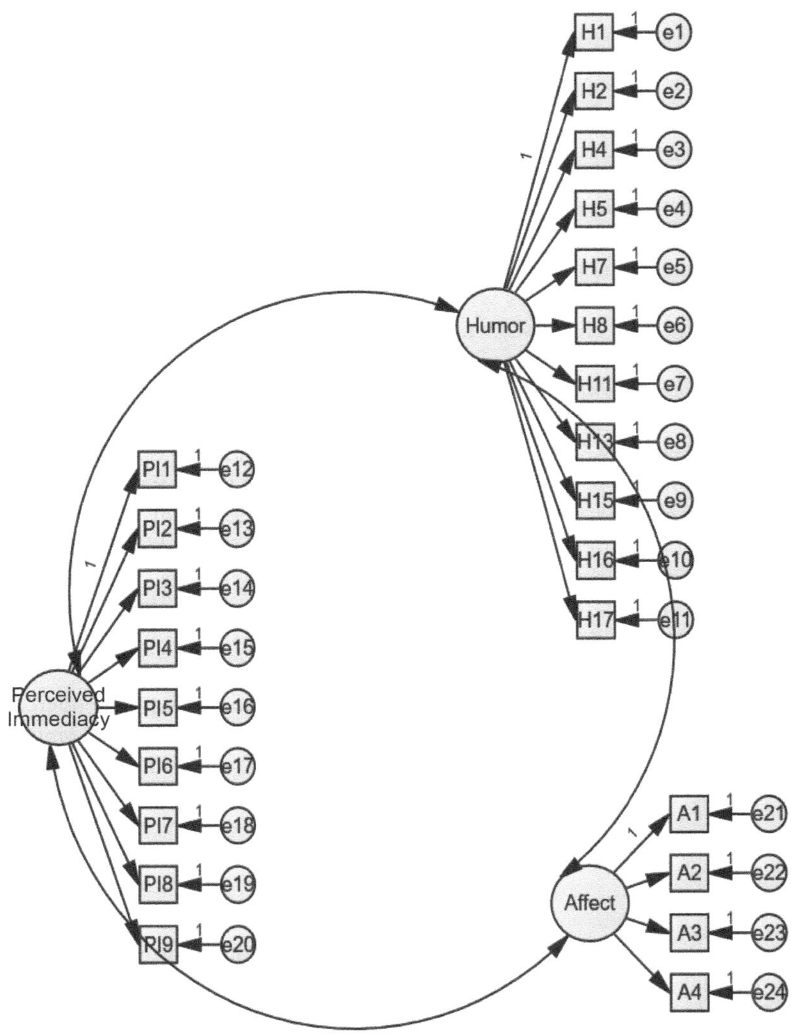

Figure 20.4 Parallelism test.

The parallelism test yields the following fit: GFI =.81, CFI =.93, RMSEA =.08, and SRMR =.05. As such, GFI and CFI indicate misfit. Table 20.8 shows the standard residual covariance matrix for all the measures combined in CFA. Note that when examining this matrix in a test of parallelism, you search not for the item that has the highest number of statistically significant residual error results, but rather the item that causes statistically significant residual error across the most measures. Only H7 arises as problematic, causing error within its own measure and the perceived immediacy measure. Thus, H7 is removed.

Recall that you knew H7 was a weak item from the test of internal consistency and factor loadings, so it is not surprising that it failed a test of parallelism, arising as a stronger assessment for a different, correlated variable. Once H7 is removed, the fit statistics are as follows: GFI =.82, CFI =.93, RMSEA =.08, and SRMR =.05. GFI is still too low, and yet no other variables emerge, causing statistically significant amounts of residual error after the multi-measure model is respecified.

From here, with no items causing statistically significant residual error on multiple measures, there are two separate philosophies about what should happen. The first philosophy is that you declare that there is too much error in the humor orientation measure for the data to guide you to a better model, and therefore it should not be used. The second philosophy of thought is that you should use the **modification indices** to find problematic items and remove them until fit is secured. Modification indices tell you how much the chi-square value will decrease if a particular relationship is removed from the model. Which is the correct course of action is up to you. Your authors recommend the more conservative option of declaring the measure inappropriate for use. However, the next few pages will show how to use the modification indices to improve fit in case this method best meets your needs or ideals. The modification indices are shown in Table 20.9.

Modification indices are displayed in terms of relationships between errors. To have the most improvement on the model, you should remove the error that has the greatest values overall associated with it. In this case, that is error 20 (e20). Looking back at Figure 20.4, you will see that error 20 is assigned to P9, the ninth item of perceived immediacy. This is precisely why the authors advise against using this method. The CFA assumes all measures input into the test of parallelism have been vetted with equal rigor, so problems caused by poor humor items may make a valid item in another measure look problematic. Nonetheless, showing how to proceed with this method, the parallelism model is respecified by removing P9. As a result, the model fit is as follows: GFI =.85, CFI =.95, RMSEA =.07, and SRMR =.05. CFI is now within an acceptable range.

Moving forward with this method, a new table of modification indices is rendered each time an item is removed. Following this method, once H2, P2, and P8 are removed, the model has acceptable fit: GFI =.90, CFI =.97, RMSEA =.06, and SRMR =.05. As such, if the humor model were to be used in hypothesis testing, only the items retained through parallelism would be utilized: items 1, 4, 5, 8, 11, 13, 15, 16, and 17.

CFA in SPSS/AMOS

To conduct a CFA, you will open a program called AMOS rather than SPSS. AMOS is an SPSS add-on program that works with SPSS to conduct structural equation modeling. Before you open AMOS, you should conduct any data cleaning assessments needed and save your prepped SPSS file.

Table 20.8 Standard Residual Covariance Matrix, Parallelism Test

	A4	*A3*	*A2*	*A1*	*P9*	*P8*	*P7*	*P6*	*P5*	*P4*	*P3*	*P2*	*P1*	*H17*	*H16*	*H15*	*H13*	*H11*	*H8*	*H7*	*H5*	*H4*	*H2*	*H1*
A4	0.00																							
A3	−0.04	0.00																						
A2	−0.09	−0.19	0.00																					
A1	−0.02	0.00	0.40	0.00																				
P9	0.20	0.29	−0.53	−0.49	0.00																			
P8	0.96	1.06	0.41	−0.02	1.08	0.00																		
P7	0.35	0.69	−0.14	−0.22	−0.30	−0.49	0.00																	
P6	0.17	−0.45	−1.06	−1.23	0.51	0.19	−0.04	0.00																
P5	0.18	0.23	−0.14	−0.70	−0.78	−0.50	0.55	0.01	0.00															
P4	0.34	0.79	−0.54	−1.02	−0.28	−0.03	0.12	−0.01	0.02	0.00														
P3	0.25	−0.03	0.30	−0.35	0.12	0.47	−0.19	0.07	0.05	−0.09	0.00													
P2	0.29	0.51	0.47	−0.36	−0.35	−0.56	0.24	−0.49	0.53	−0.03	−0.22	0.00												
P1	0.58	0.47	−0.44	−0.89	−0.26	−0.33	−0.04	−0.19	0.03	0.22	−0.18	1.17	0.00											
H17	−0.86	0.18	−0.60	−0.30	−0.54	−0.56	−0.28	−0.94	−0.56	0.14	−1.68	0.47	0.03	0.00										
H16	−0.16	0.48	−0.47	0.19	−0.63	−0.86	−0.50	−1.44	−1.35	−0.94	−1.46	0.52	−0.11	1.07	0.00									
H15	−0.45	−0.18	−0.92	−1.07	−0.01	−0.26	−0.46	−0.61	−0.82	0.63	−1.87	−0.37	−0.85	0.24	0.57	0.00								
H13	1.18	1.39	1.03	1.35	1.24	1.07	1.22	1.34	0.86	1.37	0.80	1.13	0.75	0.16	−1.83	−0.39	0.00							
H11	0.09	−0.69	−0.28	−0.28	0.82	0.82	−0.24	0.26	0.05	1.24	0.02	1.20	0.73	−0.33	−0.77	0.96	−0.77	0.00						
H8	0.62	0.36	0.74	0.63	0.56	0.40	0.04	0.35	0.19	−0.50	1.47	−0.56	0.04	−0.31	−1.40	0.26	−0.76	0.29	0.00					
H7	0.72	0.08	1.06	1.22	−1.68	−1.05	−1.48	−2.20	−1.16	−0.45	−0.80	−0.74	−1.34	−0.76	0.70	0.66	−2.81	2.40	−0.27	0.00				
H5	−0.59	0.80	−0.49	0.04	−0.36	−0.28	0.04	−0.51	−0.11	0.83	−1.57	0.51	−0.27	−0.08	−0.22	−0.25	−0.04	−0.11	−0.54	−0.17	0.00			
H4	0.48	1.13	0.47	0.99	0.79	0.78	0.68	−0.01	−0.18	0.89	−0.85	0.33	−0.18	−0.14	−0.72	0.25	0.05	0.38	0.51	0.40	0.17	0.00		
H2	0.69	1.38	−0.01	0.98	1.00	0.91	0.88	0.82	0.79	0.91	−0.67	0.24	−0.12	−0.60	−0.97	0.09	−0.10	−0.10	0.58	−0.52	−0.20	0.57	0.00	
H1	0.43	1.14	−0.28	0.33	1.35	1.17	1.39	1.22	0.58	1.92	−0.41	0.85	1.36	0.11	−0.93	−0.32	−0.21	0.10	0.32	−0.15	−0.14	−0.27	1.29	0.00

Table 20.9 Modification Indices, First Iteration

			M.I.	*Par Change*
e24	<-->	Perceived_Immediacy	4.529	.092
e21	<-->	Perceived_Immediacy	7.665	−.142
e21	<-->	e22	4.162	.118
e19	<-->	Affect	6.541	.116
e19	<-->	e20	51.478	.225
e18	<-->	e20	5.642	−.065
e18	<-->	e19	15.268	−.100
e17	<-->	Affect	7.473	−.099
e17	<-->	e24	10.713	.083
e17	<-->	e23	6.789	−.077
e17	<-->	e22	6.357	−.093
e17	<-->	e20	22.122	.116
e16	<-->	e20	34.251	−.165
e16	<-->	e19	15.017	−.102
e16	<-->	e18	25.106	.115
e15	<-->	Humor	6.462	.119
e15	<-->	e23	7.111	.100
e15	<-->	e21	6.245	−.094
e14	<-->	Humor	17.526	−.188
e14	<-->	e22	6.428	.113
e14	<-->	e19	10.142	.090
e13	<-->	e22	5.069	.123
e13	<-->	e20	4.166	−.075
e13	<-->	e19	11.243	−.116
e13	<-->	e17	17.080	−.113
e13	<-->	e16	13.004	.112
e12	<-->	e19	4.002	−.067
e12	<-->	e13	39.773	.248
e11	<-->	e13	5.656	.109
e11	<-->	e12	5.413	.103
e10	<-->	e13	6.785	.185
e10	<-->	e11	9.620	.246
e9	<-->	e21	5.137	−.110
e8	<-->	e10	6.275	−.289
e7	<-->	e23	6.561	−.154
e7	<-->	e18	9.181	−.126
e7	<-->	e9	5.738	.146
e6	<-->	e20	4.542	.086
e6	<-->	e10	12.975	−.278
e4	<-->	e24	4.743	−.080
e4	<-->	e6	7.752	−.120
e3	<-->	e6	5.016	.107
e2	<-->	e17	4.299	.065
e2	<-->	e13	6.964	−.123
e2	<-->	e12	6.499	−.115
e2	<-->	e11	6.328	−.130
e2	<-->	e10	4.622	−.174
e2	<-->	e6	4.763	.110
e2	<-->	e3	4.623	.108
e1	<-->	Perceived_Immediacy	5.388	.133
e1	<-->	e13	4.441	−.104
e1	<-->	e10	4.844	−.188
e1	<-->	e2	19.580	.247

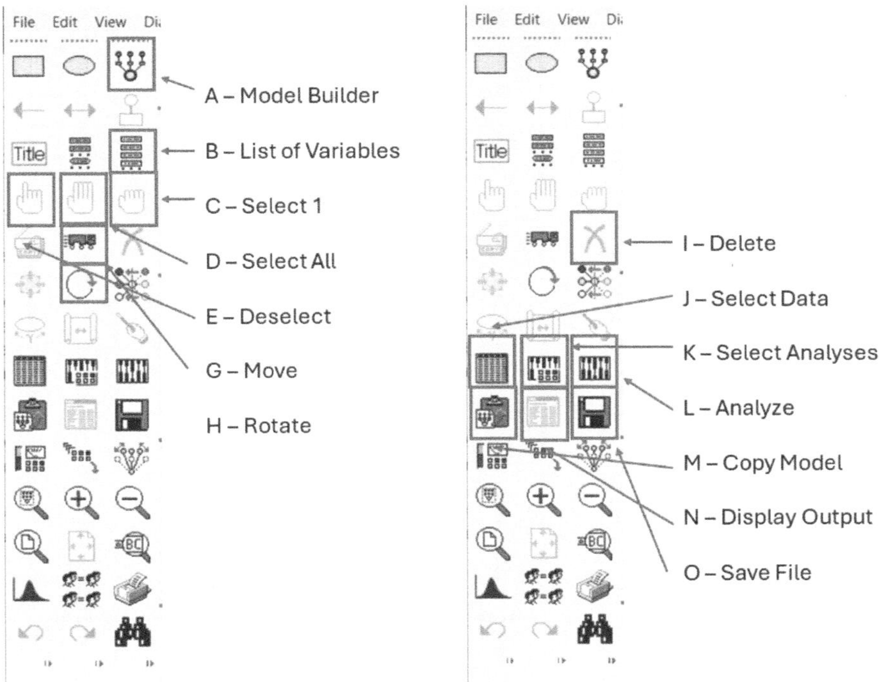

Figure 20.5 Toolbar.

Part of your data cleaning must involve resolving missing data. AMOS cannot handle missing data. You have two options for managing missing data before using AMOS:

- Delete rows of data with missing values in the latent variables. (This is advisable for any rows of data missing more than one item per latent variable measure. It is likely that there was a misunderstanding with that questionnaire, leading to so many skipped items.)
- Fill in the missing item's value with the average of all other responses in the measure. For example, if you have a 5-item measure and someone skips item 2, replace that missing value with the average response from items 1, 3, 4, and 5. This is the best option if there is just one item skipped in a measure. Be sure that if you use this option that any relevant reverse coding needed has taken place before you begin replacing missing values.

It is recommended that to use AMOS, you save an extra copy of your file and delete all labels from the Variable View. Connecting this label-free version of your data to AMOS will make your analysis easier. AMOS displays labels when they are available, which can clutter your screen, particularly if you used a program like Qualtrics to collect data that saves entire item wording as labels.

Once you open AMOS, you will see a blank canvas with the toolbar shown in Figure 20.6. You will use each of these tools at some point during your analyses.

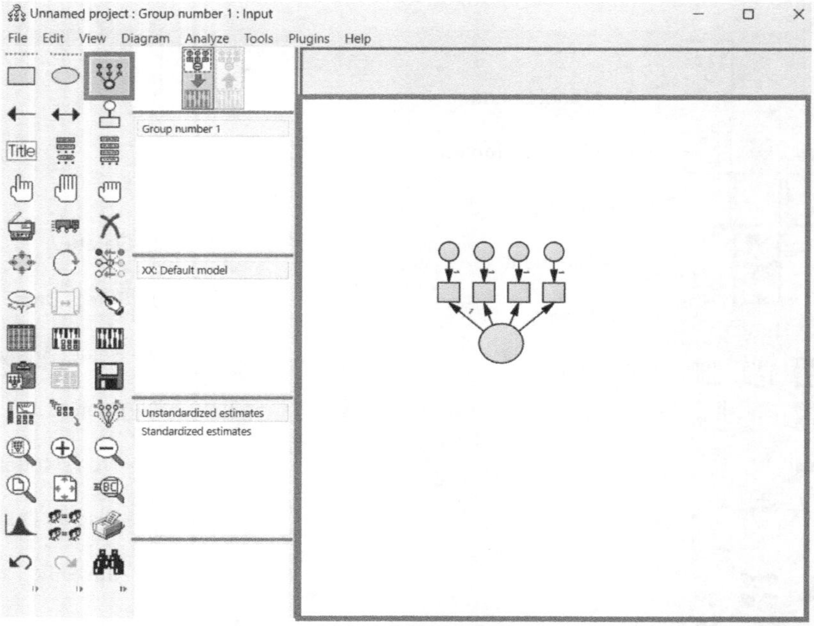

Figure 20.6 Initial model build.

The first tool you will use is tool A, the model builder. To use this tool:

1. Click on this tool (A in Figure 20.5)
2. Click anywhere on your blank canvas (the space to the right of the toolbar in Figure 20.7). Once you do this, a circle will appear. That circle is the circle that will represent your latent variable (e.g., where "Humor," "Perceived Immediacy," and "Affect" appear in Figure 20.5)
3. For every item your measure has, click on that circle again (That means that if you have a measure of four items, you select the tool, click on the canvas once for the latent variable, then four more times for the items)

Figure 20.6 shows what your canvas will look like if you click on this tool and then click the canvas in the same place five times. If you click somewhere else on the canvas, you will create a new measurement model, starting with a latent variable circle. At any time, you can add to the measurement model by clicking the Model Builder button and then clicking the latent variable circle. Once your measure has enough items, click on the Model Builder again to deactivate it.

Just as your initial circle represents a specific piece of your model, so do the other objects. The boxes represent the items from your measure. The little circles represent the error acting upon each item (e.g., wording problems, sampling error, psychological noise).

Next, you need to tell AMOS what goes into each of the pieces of your model. Before you begin adding pieces, it is highly recommended that you:

4. Click on the Rotate tool (H in Figure 20.5)
5. Click on the latent variable circle

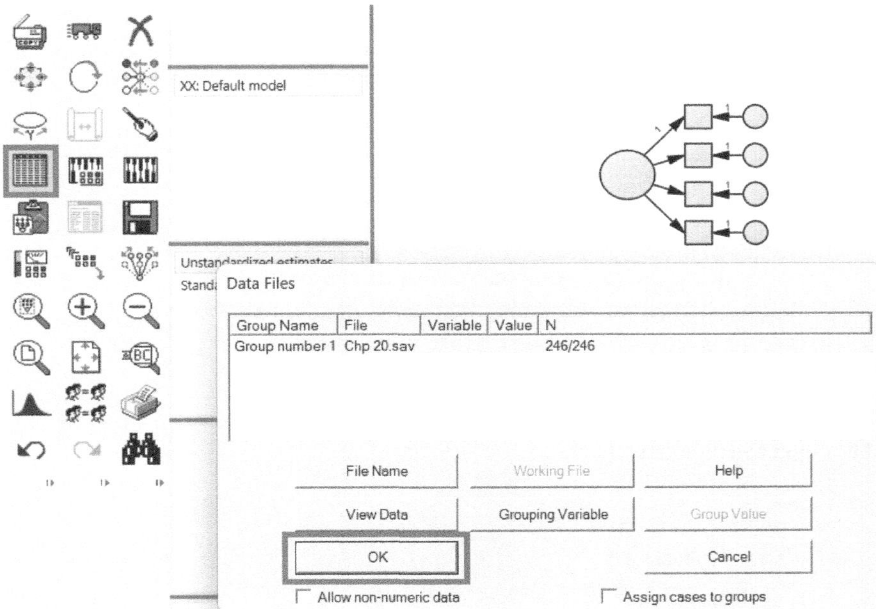

Figure 20.7 Connect to data.

Doing this will rotate the model 90°, as shown in Figure 20.7. Rotating the model like this will make it easier to read your model content later if your item names have more than two characters. Next, you will connect your data to your model:

6. Click the Select Data button (J in Figure 20.5). This will open the box shown in Figure 20.7
7. Navigate to where your data is saved and select it. Remember, it is recommended that you use a copy of your SPSS file that has the labels removed
8. Click OK

Now you have connected to your data. It is time to use that data to build your model:

9. Double click on your latent variable. This will open the Object Properties box shown in Figure 20.9
10. Under Variable name, type the name of your variable without spaces. Like SPSS, AMOS cannot handle spaces. So, while you can type "Humor_Orientation" here, "Humor Orientation" will give you an illegal character error message. If you wish, you may give your variable a label in this same box. For example, here is where you can list the actual name of the variable as "Humor Orientation"
11. Click X to close the box

When it comes to labeling your variables, should you choose to engage in this step, briefer is better. Because AMOS uses a visual model builder, anything you type in will appear in the model. This means text sometimes overlaps when you have long titles. In the case of the humor orientation

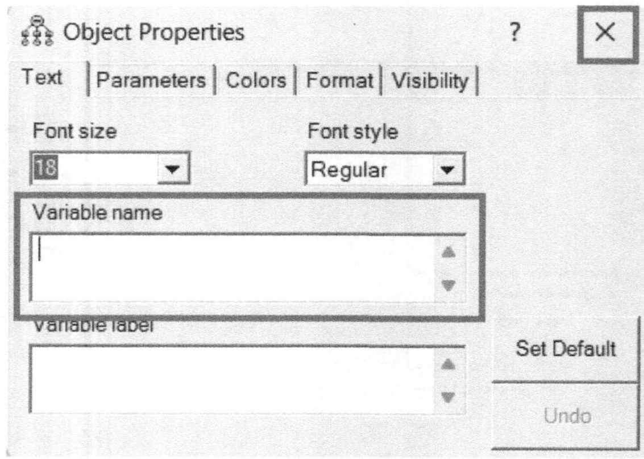

Figure 20.8 Object properties.

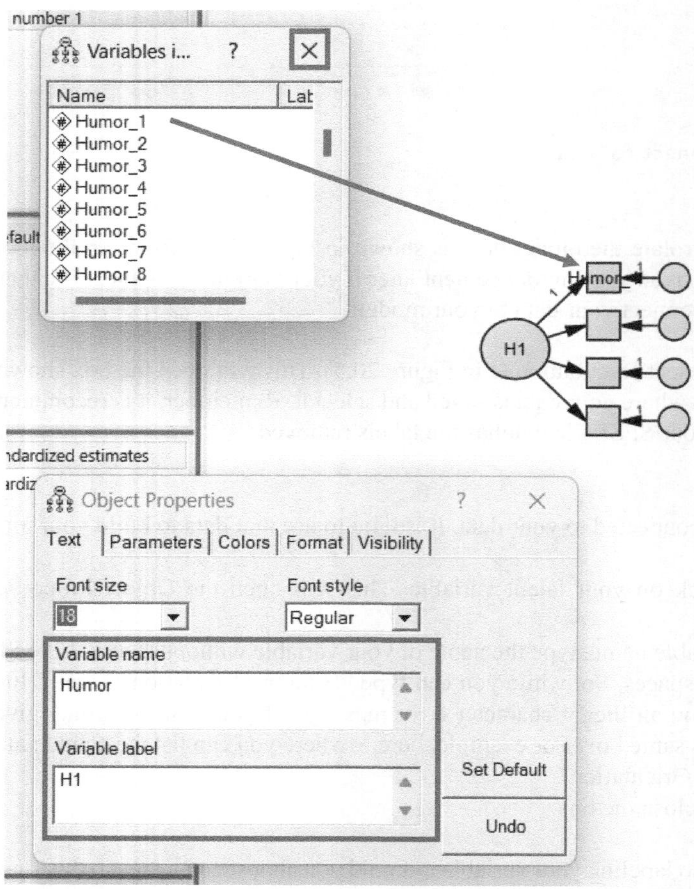

Figure 20.9 List of variables.

measure shown in Figure 20.1, the measure was given the short label "Humor" and each item was given a short label "H" plus a number.

Next, individual items will be identified:

12. Click on the List of Variables tool (B in Figure 20.5)
13. Click on the items in your measure one at a time, dragging a single measure item to each box

Figure 20.9 shows the list of variables that opens once the button is selected, as well as what the measurement model looks like once an item has been moved to the first available box. You can choose to click the X to close the List of Variables at this time or leave it open.

Once all boxes are connected to the items they represent, the error circles should be filled. You can do this manually by clicking on each circle and giving it a name, but there is a much quicker way to do this. After all items in your model's items are connected to the data, use the plugin to label errors. As shown in Figure 20.10:

14. Click on the Plugin menu
15. Click Name Unobserved Variables

This fills all error terms automatically.

At this point, the model is fully built. Sometimes after everything is filled in, you will find yourself wishing that you had built somewhere else in the workspace or could spread the model out. Tool C (Select 1) in Figure 20.5 will allow you to select individual pieces of the model by clicking that tool and then clicking the pieces. Any selected pieces will turn blue. Once these pieces turn blue, they are grouped and selected. You can move these pieces by clicking the Move tool (G in Figure 20.5) and then dragging and dropping your selected blue pieces. Once the pieces are in place, click the Deselect button (E in Figure 20.5). You can also select every piece of the model at once to move by clicking the Select All button (D in Figure 20.5) and then utilizing the Move tool. (Note that if you have more than one model drawn and use Select All, everything in your build space will be selected.) Like when you use the Select 1 tool, you should click the Deselect button once your model is relocated. Notably, you can use these selecting, deselecting, and moving tools at any point in a build. They are mentioned here as a common point of need.

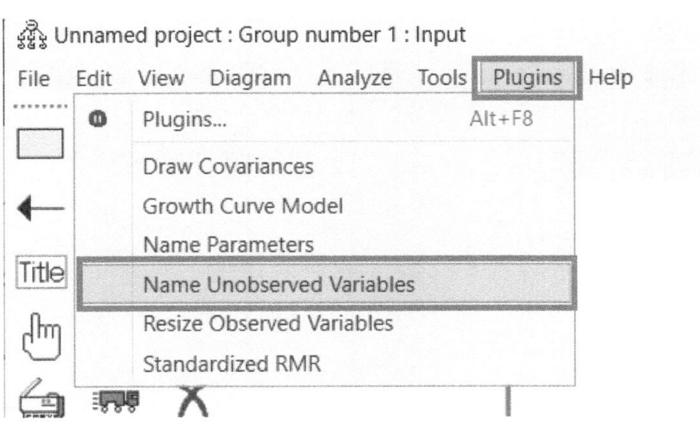

Figure 20.10 Name Unobserved Variables.

Now it is time to start the analysis:

16. Click on Select Analyses (K in Figure 20.5). This opens the Analysis Properties window (Figure 20.11)
17. Click on the output tab
18. Click the boxes beside Minimization history, Standardized estimates, Residual moments, and Modification indices to ensure they are selected with a checkmark
19. Click X to close the window

At this point, you have built your model and told AMOS almost everything you want to be output from its list of standard offerings.

As a side note, if you click on the Estimation tab shown in Figure 20.11, it allows you to change the parameter estimation algorithm driving your CFA. The default is Maximum Likelihood, which

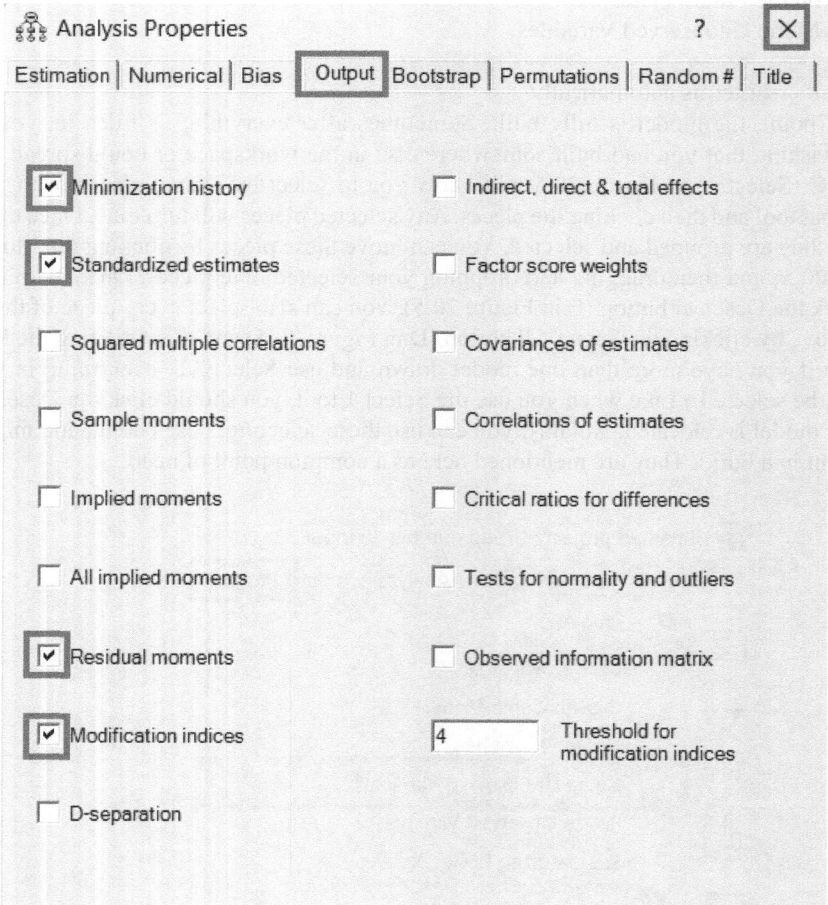

Figure 20.11 Analysis properties.

assumes that each item in the measure has a normal distribution. Your items should always be fairly normally distributed, and if they are not, that's a sign that something is amiss. However, if your items are not normally distributed, you should use Unweighted Least Squares rather than Maximum Likelihood. When using this parameter estimation algorithm, you will need to rely on different fit statistics (Adjusted CFI and χ^2) along with SRMR. We recommend reading Kline (2023) to better understand how to handle this unusual circumstance.

The last step of output setup is to tell AMOS that you also want the SRMR produced. SRMR is one of the more recently developed fit statistics, so it was not part of the original AMOS program. (You can think of the Plugin menu in AMOS as the menu of tools that researchers have communicated a need for since AMOS was originally developed.) To tell AMOS you want the SRMR:

20. Click the Plugin menu
21. Click Standard RMR

Now you have told AMOS absolutely everything it needs to know to analyze the data.

To analyze the data, AMOS will ask you to first save your file. This step is necessary to free RAM before the labor-intensive analyses begin for your computer. The steps to analyzing the data are:

22. Click the Save button (O in Figure 20.5). AMOS will produce and save several files for the model, so it is recommended that you save your model to a folder rather than generally on the Desktop or in Documents
23. Click Analyze (L in Figure 20.5)

At this point, your SRMR will appear in a small window of its own. All other data will appear in a large output window. If at any time you accidentally close either of those windows, redo the steps shown in Figure 20.12 to open the SRMR window. Click on the Display Output button (N in Figure 20.5) to reopen the reset of the output.

The output window is shown in Figure 20.13. GFI, CFI, and RMSEA are available in the Model Fit tab. Modification Indices are shown in the tab of the same name. The Standard Residual Covariance matrix appears at the bottom of the Estimates tab. See Figure 20.13 for the location of these tabs.

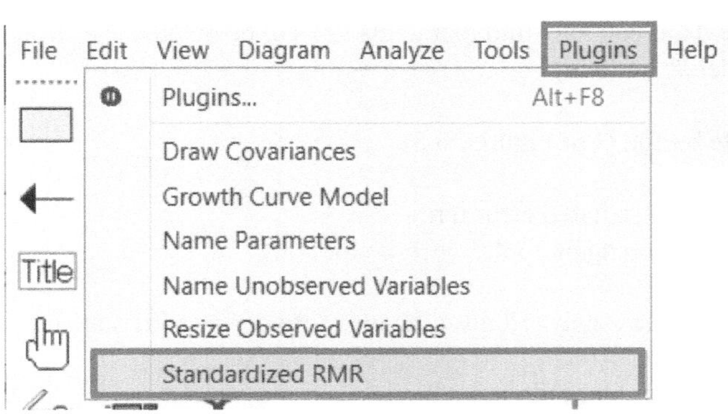

Figure 20.12 SRMR.

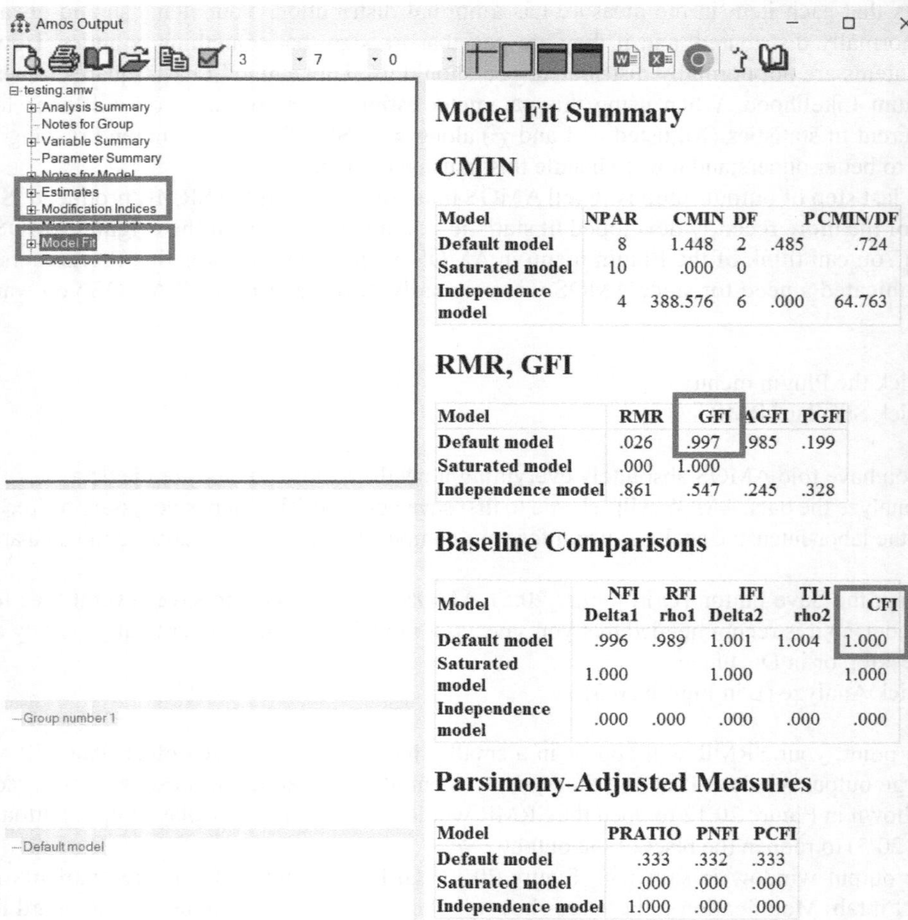

Model Fit Summary

CMIN

Model	NPAR	CMIN	DF	P	CMIN/DF
Default model	8	1.448	2	.485	.724
Saturated model	10	.000	0		
Independence model	4	388.576	6	.000	64.763

RMR, GFI

Model	RMR	GFI	AGFI	PGFI
Default model	.026	.997	.985	.199
Saturated model	.000	1.000		
Independence model	.861	.547	.245	.328

Baseline Comparisons

Model	NFI Delta1	RFI rho1	IFI Delta2	TLI rho2	CFI
Default model	.996	.989	1.001	1.004	1.000
Saturated model	1.000		1.000		1.000
Independence model	.000	.000	.000	.000	.000

Parsimony-Adjusted Measures

Model	PRATIO	PNFI	PCFI
Default model	.333	.332	.333
Saturated model	.000	.000	.000
Independence model	1.000	.000	.000

Figure 20.13 Output.

After examining the data, you may find that items need to be deleted (i.e., you may need to respecify the model). To delete an item:

24. Click the Delete button (I of Figure 20.5)
25. Click on the item
26. Click on the item's associated error term
27. Click the Delete button again to deselect the tool

After that, you will hit the Analyze button again to render new fit statistics and a new standard residual covariance matrix.

Once you have your final measurement model, you may find that your fit statistics are unstandardized (i.e., not between −1 and 1). To fix this, click Standardized estimates, as shown in Figure 20.14.

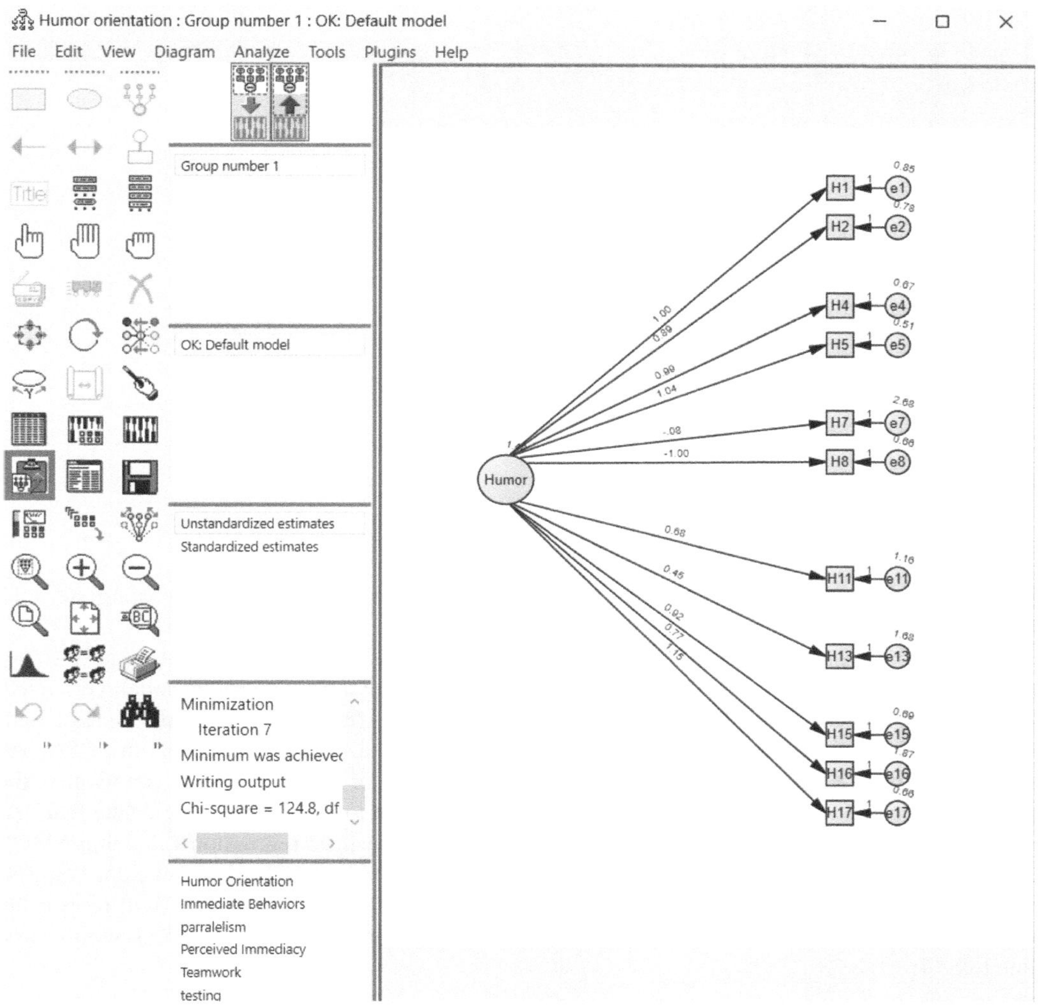

Figure 20.14 Other tools.

You may also wish to copy your measurement model into a paper. To make a copy of the model, click the Copy Model button (M from Figure 20.5).

The last tool you are likely to need is the button that connects measurement models in tests of parallelism. Up to this point, everything you have been shown are tools you will use in all internal consistency tests. If you look back at Figure 20.4, you'll see that all latent variables are connected with a double arrow. To make these connections:

28. Click the double arrow indicated in Figure 20.14
29. Click on one of your latent variables
30. Click on the variable you wish to connect the first variable to with a double arrow

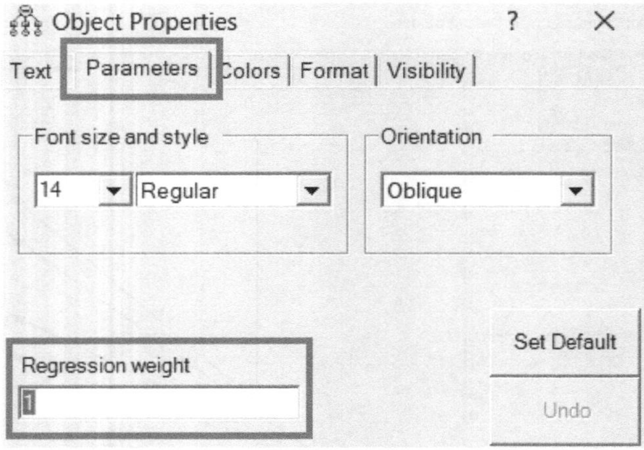

Figure 20.15 Regression weight.

Once all latent variables are connected, you can run tests of parallelism by repeating steps 17 through 28 on the connected models.

There is one last thing you may need to know about AMOS to run a CFA. This is only something you would need to know if, in the process of respecifying a measurement model, you delete the first item. Notice that in all figures in this chapter that show a measurement model, the first item of each measure has a 1 next to it, even before analyzing the data. This is because to run a CFA, one of the items must be standardized to have a regression weight of 1 so that AMOS can identify the potential range of variables and standardize them with respect to that item. If you delete that first item in any measurement model, you have removed the standardized regression value that AMOS set for you. That means you'll need to give another item that standardized beta weight value. If you right click on another line running from the latent variable to the item, that will reopen the Object Properties window from Figure 20.8. To assign the regression weight represented in that line to be 1:

31. Click on the Parameters tab of Object Properties as shown in Figure 20.15
32. Type "1" into the Regression weight box
33. Click X to close the window

That will allow you to continue conducting CFAs on the measure with intelligible output.

Summary

CFA is a method of determining whether a measurement model behaves the way it theoretically should when used in your sample. You should use at least a test of internal consistency of measures through CFA any time you use such assessments in quantitative research. It is not advisable to start building your own measures, relying on CFA alone for evidence of validity, when you need to develop assessments from scratch. Take a measurement class first, or work with a psychometrician. There is much more to measurement development than shown here. Lego Dumbledore would need more than CFA and a magical wand to ensure his measure is developed well. If you are interested in measurement development work to connect your statistical knowledge with your knowledge in your chosen field, schools that offer such training typically house their psychometrics courses in psychology or education.

References

Bandalos, D. L. (2018). *Measurement theory and applications for the social sciences*. The Guilford Press.

Bolkan, S., & Goodboy, A. K. (2015). Exploratory theoretical tests of the instructor humor–student learning link. *Communication Education*, *64*(1), 45–64. https://doi.org/10.1080/03634523.2014.978793

Booth-Butterfield, S., & Booth-Butterfield, M. (1991). Individual differences in the communication of humorous messages. *Southern Journal of Communication*, *56*(3), 205–218. https://doi.org/10.1080/10417949109372831

Kelly, S., Rice, C., Wyatt, B., Ducking, J., & Denton, Z. (2015). Teacher immediacy and decreased student quantitative reasoning anxiety: The mediating effect of perception. *Communication Education*, *64*(2), 171–186. https://doi.org/10.1080/03634523.2015.1014383

Kelly, S., & Westerman, D. K. (2020). Doing communication science: Thoughts on making more valid claims. *Annals of the International Communication Association*, *44*(3), 177–184. https://doi.org/10.1080/23808985.2020.1792789

Kelly, S., & Westerman, D. K. (2024). *A primer in the philosophy of science: A guide to thinking like a scientist for social scientists*. Kendall Hunt.

Kline, R. B. (2023). *Principles and practice of structural equation modeling* (5th ed.). Guilford Press.

McCroskey, J. C. (1994). Assessment of affect toward communication and affect toward instruction in communication. In S. Morreale & M. Brooks (Eds.), *1994 SCA summer conference proceedings and prepared remarks: Assessing college student competence in speech communication*. Annandale, VA: Speech Communication Association.

Violanti, M. T., Kelly, S. E., Garland, M. E., & Christen, S. (2018). Instructor clarity, humor, immediacy, and student learning: Replication and extension. *Communication Studies*, *69*(3), 251–262. https://doi.org/10.1080/10510974.2018.1466718

Appendix T

Critical Value of *t*

df	*0.1*	*0.05*	*0.025*	*0.01*	*0.005*	*0.001*
1	3.078	6.314	12.706	31.821	63.657	318.310
2	1.886	2.920	4.303	6.965	9.925	22.327
3	1.638	2.353	3.182	4.541	5.841	10.214
4	1.533	2.132	2.776	3.747	4.604	7.173
5	1.476	2.015	2.571	3.365	4.032	6.869
6	1.440	1.943	2.447	3.143	3.707	5.959
7	1.415	1.895	2.365	2.998	3.499	5.408
8	1.397	1.860	2.306	2.896	3.355	5.041
9	1.383	1.833	2.262	2.821	3.250	4.781
10	1.372	1.812	2.228	2.764	3.169	4.587
11	1.363	1.796	2.201	2.718	3.106	4.437
12	1.356	1.782	2.179	2.681	3.055	4.318
13	1.350	1.771	2.160	2.650	3.012	4.221
14	1.345	1.761	2.145	2.624	2.977	4.140
15	1.341	1.753	2.131	2.602	2.947	4.073
16	1.337	1.746	2.120	2.583	2.921	4.015
17	1.333	1.740	2.110	2.567	2.898	3.965
18	1.330	1.734	2.101	2.552	2.878	3.922
19	1.328	1.729	2.093	2.539	2.861	3.883
20	1.325	1.725	2.086	2.528	2.845	3.850
21	1.323	1.721	2.080	2.518	2.831	3.819
22	1.321	1.717	2.074	2.508	2.819	3.792
23	1.319	1.714	2.069	2.500	2.807	3.767
24	1.318	1.711	2.064	2.492	2.797	3.745
25	1.316	1.708	2.060	2.485	2.787	3.725
30	1.310	1.697	2.042	2.457	2.750	3.646
40	1.303	1.684	2.021	2.423	2.704	3.551
60	1.296	1.671	2.000	2.390	2.660	3.460
100	1.290	1.660	1.984	2.364	2.626	3.390
120	1.289	1.658	1.980	2.358	2.617	3.373
150	1.287	1.655	1.976	2.350	2.609	3.357
200	1.286	1.652	1.972	2.345	2.601	3.340
300	1.284	1.650	1.968	2.340	2.597	3.319
500	1.283	1.648	1.965	2.337	2.593	3.310
1000	1.282	1.646	1.962	2.334	2.589	3.300

For any particular degrees of freedom, entry represents the critical value of *t* corresponding to the cumulative probability (1- α) and a specified upper-tail area (α)

Appendix Z

Z Cumulative Distribution Table

z	*0*	*0.01*	*0.02*	*0.03*	*0.04*	*0.05*	*0.06*	*0.07*	*0.08*	*0.09*
0	0.5000	0.5040	0.5080	0.5120	0.5160	0.5199	0.5239	0.5279	0.5319	0.5359
0.1	0.5398	0.5438	0.5478	0.5517	0.5557	0.5596	0.5636	0.5675	0.5714	0.5753
0.2	0.5793	0.5832	0.5871	0.5910	0.5948	0.5987	0.6026	0.6064	0.6103	0.6141
0.3	0.6179	0.6217	0.6255	0.6293	0.6331	0.6368	0.6406	0.6443	0.6480	0.6517
0.4	0.6554	0.6591	0.6628	0.6664	0.6700	0.6736	0.6772	0.6808	0.6844	0.6879
0.5	0.6915	0.6950	0.6985	0.7019	0.7054	0.7088	0.7123	0.7157	0.7190	0.7224
0.6	0.7257	0.7291	0.7324	0.7357	0.7389	0.7422	0.7454	0.7486	0.7517	0.7549
0.7	0.7580	0.7611	0.7642	0.7673	0.7704	0.7734	0.7764	0.7794	0.7823	0.7852
0.8	0.7881	0.7910	0.7939	0.7967	0.7995	0.8023	0.8051	0.8078	0.8106	0.8133
0.9	0.8159	0.8186	0.8212	0.8238	0.8264	0.8289	0.8315	0.8340	0.8365	0.8389
1	0.8413	0.8438	0.8461	0.8485	0.8508	0.8531	0.8554	0.8577	0.8599	0.8621
1.1	0.8643	0.8665	0.8686	0.8708	0.8729	0.8749	0.8770	0.8790	0.8810	0.8830
1.2	0.8849	0.8869	0.8888	0.8907	0.8925	0.8944	0.8962	0.8980	0.8997	0.9015
1.3	0.9032	0.9049	0.9066	0.9082	0.9099	0.9115	0.9131	0.9147	0.9162	0.9177
1.4	0.9192	0.9207	0.9222	0.9236	0.9251	0.9265	0.9279	0.9292	0.9306	0.9319
1.5	0.9332	0.9345	0.9357	0.9370	0.9382	0.9394	0.9406	0.9418	0.9429	0.9441
1.6	0.9452	0.9463	0.9474	0.9484	0.9495	0.9505	0.9515	0.9525	0.9535	0.9545
1.7	0.9554	0.9564	0.9573	0.9582	0.9591	0.9599	0.9608	0.9616	0.9625	0.9633
1.8	0.9641	0.9649	0.9656	0.9664	0.9671	0.9678	0.9686	0.9693	0.9699	0.9706
1.9	0.9713	0.9719	0.9726	0.9732	0.9738	0.9744	0.9750	0.9756	0.9761	0.9767
2	0.9772	0.9778	0.9783	0.9788	0.9793	0.9798	0.9803	0.9808	0.9812	0.9817
2.1	0.9641	0.9656	0.9671	0.9686	0.9699	0.9713	0.9726	0.9738	0.9750	0.9761
2.2	0.9772	0.9783	0.9793	0.9803	0.9812	0.9821	0.9829	0.9837	0.9845	0.9852
2.3	0.9859	0.9866	0.9872	0.9878	0.9884	0.9889	0.9894	0.9898	0.9902	0.9906
2.4	0.9910	0.9913	0.9916	0.9918	0.9920	0.9922	0.9925	0.9927	0.9929	0.9931
2.5	0.9932	0.9934	0.9935	0.9936	0.9938	0.9939	0.9940	0.9941	0.9942	0.9943
2.6	0.9944	0.9945	0.9946	0.9947	0.9948	0.9949	0.9950	0.9951	0.9952	0.9953
2.7	0.9953	0.9954	0.9955	0.9956	0.9956	0.9957	0.9958	0.9959	0.9959	0.9960
2.8	0.9961	0.9961	0.9962	0.9962	0.9963	0.9964	0.9964	0.9965	0.9965	0.9966
2.9	0.9966	0.9967	0.9967	0.9968	0.9968	0.9969	0.9969	0.9970	0.9970	0.9971
3	0.9971	0.9972	0.9972	0.9973	0.9973	0.9973	0.9974	0.9974	0.9974	0.9975
3.5	0.9990	0.9990	0.9991	0.9991	0.9991	0.9992	0.9992	0.9992	0.9993	0.9993
4	0.9997	0.9997	0.9997	0.9997	0.9997	0.9997	0.9997	0.9998	0.9998	0.9998
4.5	0.9999	0.9999	0.9999	0.9999	0.9999	0.9999	0.9999	0.9999	0.9999	0.9999
5	1.0000	1.0000	1.0000	1.0000	1.0000	1.0000	1.0000	1.0000	1.0000	1.0000

This table provides the proportion of the normal distribution less than or equal to the *z*-score.

Author Index

Subject Index

For Product Safety Concerns and Information please contact our EU representative GPSR@taylorandfrancis.com Taylor & Francis Verlag GmbH, Kaufingerstraße 24, 80331 München, Germany